FERRARI
250 GT
BERLINETTA
"Tour de France"

Other Veloce publications -

SpeedPro Series
How to Blueprint & Build a 4-Cylinder Engine Short Block for High Performance by Des Hammill
How to Build a V8 Engine Short Block for High Performance by Des Hammill
How to Build & Power Tune Weber DCOE & Dellorto DHLA Carburetors by Des Hammill
How to Build & Power Tune Harley-Davidson Evolution Engines by Des Hammill
How to Build & Power Tune Distributor-type Ignition Systems by Des Hammill
How to Build, Modify & Power Tune Cylinder Heads by Peter Burgess
How to give your MGB V8 Power by Roger Williams
How to Power Tune the MGB 4-Cylinder Engine by Peter Burgess
How to Power Tune the MG Midget & Austin-Healey Sprite by Daniel Stapleton
How to Power Tune Alfa Romeo Twin Cam Engines by Jim Kartalamakis
How to Power Tune Ford SOHC 'Pinto' & Sierra Cosworth DOHC Engines by Des Hammill

Colour Family Album Series
Bubblecars & Microcars by Andrea & David Sparrow
Bubblecars & Microcars, More by Andrea & David Sparrow
Citroen 2CV by Andrea & David Sparrow
Citroen DS by Andrea & David Sparrow
Lambretta by Andrea & David Sparrow
Mini & Mini Cooper by Andrea & David Sparrow
Vespa by Andrea & David Sparrow
VW Beetle by Andrea & David Sparrow
VW Bus, Camper, Van & Pick-up by Andrea & David Sparrow

General
Alfa Romeo Owner's Bible by Pat Braden
Alfa Romeo Modello 8C 2300 by Angela Cherrett
Alfa Romeo Giulia Coupé GT & GTA by John Tipler
Bugatti 46/50 - The Big Bugattis by Barrie Price
Bugatti 57 - The Last French Bugatti by Barrie Price
Chrysler 300 - America's Most Powerful Car by Robert Ackerson
Cobra - The Real Thing! by Trevor Legate
Daimler SP250 'Dart' by Brian Long
Fiat & Abarth 124 Spider & Coupé by John Tipler
Fiat & Abarth 500 & 600 by Malcolm Bobbitt
Ford F100/F150 Pick-up by Robert Ackerson
Lola History (1957-1977) by John Starkey
Lola T70 - The Racing History & Individual Chassis Record New Edition by John Starkey
Making MGs by John Price Williams
Mazda MX5/Miata Enthusiast's Workshop Manual by Rod Grainger & Pete Shoemark
MGA by John Price Williams
Mini Cooper - The Real Thing! by John Tipler
Nuvolari: When Nuvolari Raced ... by Valerio Moretti
Porsche 356 by Brian Long
Porsche 911R, RS & RSR by John Starkey
Porsche 914 & 914/6 by Brian Long
Rolls-Royce Silver Shadow/Bentley T-Series Corniche & Camargue by Malcolm Bobbitt
Rolls-Royce Silver Wraith, Dawn & Cloud/Bentley R & S Series by Martyn Nutland
Singer Story: Cars, Commercial Vehicles, Bicycles & Motorcycles by Kevin Atkinson
Triumph TR6 by William Kimberley
Triumph Motorcycles & the Meriden Factory by Hughie Hancox
Volkswagen Karmann Ghia by Malcolm Bobbitt
VW Beetle - Rise from the Ashes of War by Simon Parkinson
VW Bus Type 2 Transporter by Malcolm Bobbitt

First published in 1997 by Veloce Publishing Plc., 33, Trinity Street, Dorchester DT1 1TT, England. Fax: 01305 268864.

ISBN: 1 874105 75 8

Readers with ideas for automotive books, or books on other transport or related hobby subjects, are invited to write to Veloce Publishing at the above address.

British Library Cataloguing in Publication Data -
A catalogue record for this book is available from the British Library.

Typesetting (Bookman), design and page make-up all by Veloce on AppleMac.

Printed and bound by The Cromwell Press, Broughton Gifford, Melksham, Wiltshire

FERRARI 250 GT BERLINETTA

"Tour de France"

JOHN STARKEY

VELOCE PUBLISHING PLC
PUBLISHERS OF FINE AUTOMOTIVE BOOKS

250 GT Berlinetta

ACKNOWLEDGEMENTS

Car books are really about pictures. So I've tried in this particular book to give a truly representative spread in the photographs used. You'll find in the following pages that the cars shown range from original prototypes through competition heydays to the cars as they are today.

To assemble such a selection I have the kindness of many people to acknowledge, especially my good friend Ed Niles. I must also thank Alexis Callier and Andre Van Bever for most of the photos of the cars when they were raced. Amongst other people who were kind enough to send me photos were: the late Rob Merrill, Paul Kunkel, Dr M. Kiener and Nico Koel. Rudi Pas supplied the photos of 0415GT as she was in 1986. Chuck Weber, Chester Bolin and Bradley Balles invited me to photograph their cars.

Also my thanks to Andreas Birner and Gregor Schulz for bringing me up to date with cars which they knew about. Lance Hill gave a lot of time plus contributing his photographs and expertise over the restoration of his own car, 0707 GT. A very big thank you to everyone concerned, and my apologies to anyone I have neglected to mention.

John Starkey

250 GT *Berlinetta* FERRARI

CONTENTS

250 GT Berlinetta

FOREWORD

I was excited and delighted when I first read the manuscript of this book. I know you will have that same pleasant experience.

With the civilized world being inundated with Ferrari books, it's not unfair to ask "Why another book on Ferraris?" A more appropriate question might be "What sets this book apart from the other books on Ferrari?" To me, the answer is twofold: The precise subject matter and the enthusiasm of the author.

Let's talk about the author first. I first met John Starkey as a "pen pal" when he let it be known to the world that he was searching for a certain 250GT Ferrari Berlinetta, chassis number 0911 GT. As it happened, that very car showed up in the classified ads at the very same time and (not being above trying to put a deal together - I was in Los Angeles, the seller in Arizona), the deal was done. When the car finally arrived at it's new home, I heard from John that he was delighted with the car and pleased that it was exactly as represented. John and I were still half a world apart and had never met face to face, so I supposed that I would never hear from him again.

Not so.

Not too many months later, I received a call from John: "Hullo, Ed. I'm in town with my wife, Sandra, and I wonder if we might take you to dinner?" Thinking that we would be in for an evening of boring questions, the answers to which would generally be "Who cares?" or, worse yet, a long list of complaints about the car, I suggested a *very* expensive restaurant which my wife and I had been hoping to visit.

Well! John and Sandra Starkey were the exact opposite of boring, whatever that is! Adjectives such as delightful, charming, witty, enthusiastic, come to mind. Needless to say, that restaurant became one of our favourites and John and Sandra Starkey became our dear and true friends. As a measure of John's character, he's even forgiven me for the restaurant tab!

I used the word "enthusiastic." It's John Starkey's enthusiasm which sets this book apart from others of its ilk. There are many professional writers who have written about Ferraris, but few of them write from a position of enthusiastic ownership as does this author. John's education about Ferraris generally, and the so-called "Tour de France" in particular, has been hard earned: hard driving and hard cash at the overhaul bench. John's Ferrari was not a showpiece to be admired but never touched; on the contrary, he used it, and used it the way it was meant to be used, at every opportunity. John entered the car successfully on three separate occasions in the retrospective Mille Miglia event through Italy and the more he used the car the more his enthusiasm and respect for the model grew. John once told me that he would sell his house before he would sell his Ferrari! [This came true! *Author*].

Now to the second element of the book that makes it so exciting: the Ferrari 250GT. I've owned over one hundred Ferraris (not bragging, just establishing my credentials) ranging from 166MMs to 308s and I still find the early 250GT's the most exciting.

I wish I could say why. Heaven knows, the later cars are better insulated, smoother to drive, and infinitely more civilized. And certain cars that I've tried have been individually more exciting - the 206SP, the 375MM and the Le Mans class-winning Comp. SWB 250GT come to mind - but, overall, the long wheelbase 250GT Berlinettas are somehow the most exciting. The small-block V12 engine had just about reached its peak of development, and the cars were genuine dual-purpose vehicles capable of being driven to the market or to a class win in an international racing event. Ferrari was still clinging to the use of the drum brakes, stubbornly resisting the inexorable march of progress. The cars were winning races because of their strength and because of the ongoing development of tried and true principles, rather than by the use of innovative and experimental methods.

I owned most of my long wheelbase 250GTs in the 1960s, and it was terribly exciting to learn the racing history of a particular car - sometimes after it had been handed over to the next owner. Almost all of these cars had an interesting racing life before they showed up on the used car market; perhaps that's part of their intrigue.

One hears so much about the "Ferrari mystique." Actually, the cars themselves are very straightforward from an engineering standpoint, having been developed with regular and rapid minor changes as a result of the lessons of the racing circuits. These cars are, indeed, true examples of that old cliche "Racing improves the breed." So, perhaps, it is the very multiplicity of models resulting from these changes, together with the interesting racing histories, which generates that Ferrari mystique.

Nowhere is that mystique more evident than in the 250GT Berlinetta of 1954 to 1959.

So read this book enthusiastically, knowing that it was written by a true enthusiast!

Ed Niles

7

250 GT Berlinetta

INTRODUCTION

I stood in a daze outside my house and listened to the slam of sound as the black Ferrari accelerated past and sped away from me. I had just been given the ride of my life, and little has approached it since.

That ride took place 30 years ago, when I was 19; up to that tender age I was not interested in cars at all - a fact my car-loving friends could not understand, especially as my father had owned a string of desirable cars such as a Healey Silverstone ("Er, it's a production car, dear." - or at least that's what he told my Mother before she saw it!), MkV and MkVII Jaguars, plus assorted Sunbeams and Rovers. One evening, whilst staying at my friend Allen's house for the weekend, the two of us made our way down to his local for an evening's drinking: my friend's eyes were suddenly taken by a black car in the car park and he shot off to investigate.

Curious, I strolled over to where he was standing engrossed by the car. "Know what this is?" he enquired, without taking his eyes from the car. "No, what is it?" I replied, somewhat bored and thinking of the drinking time we were missing. "It's a Ferrari," he announced triumphantly, "a real beauty, too."

In the bar we asked who the Ferrari belonged to, and were told it was to a character by the name of Paul Kay. We watched him walk to his car, start it up and drive off down the road. My friend turned round with glazed eyes - "Did you hear that?" he said, "he was doing 60 at least - and still in first!" Later that week we encountered the car and its owner at the same pub, and this time my friend was cheeky enough to ask - successfully! - for a ride.

I tried to follow them in Allen's Ford Thames van, which managed to run out of petrol during the chase. Fortunately, Paul Kay and Allen soon returned in the Ferrari (having had enough time to reach another pub and have a drink!). On hearing of the van's problem, Paul Kay reached into the car and produced a half pint glass - kept for emergencies, he said - and proceeded to dip it into the Ferrari's large filler neck. We were amazed. "Holds 40 gallons," he explained, "I shan't miss a bit." Allen took over his van and I was lucky enough to be given a lift home in the Ferrari (about 15 miles) as Paul Kay just happened to live near me.

If you own a Ferrari, particularly one of the older competition cars, you'll have a good idea what the 15 miles were like, especially in the sixties when there were few speed limits. Paul Kay said we touched 145mph late that night on a normal two-way road, and I believe it!

That was it! My life was changed ... From then, on my overriding ambition was to own a Ferrari - not just any Ferrari - that Ferrari! Paul Kay offered it to me three months later, for £1750 - I didn't have 1750 beans. Strangely enough, I met another Ferrari enthusiast, Mark Rigg, two years later in the same part of the world. He offered me his "Tour de France" (for such I had found the type to be) chassis number 0597 GT for £1000. I did not have that either.

And now comes the strangest twist. Fifteen years went by, and I had made some money. I remembered that the 'Paul Kay' Ferrari had won the 1958 Mille Miglia, and discovered its chassis number - 0911 GT. I wrote to the English Ferrari Owner's Club and asked its whereabouts. Two weeks after the letter was printed, I received a reply from Ed Niles in Los Angeles, telling me of the car's history since it went to the States in 1974. I wrote back, thanking him for the information, and adding, "If the car ever comes up for sale, please let me know."

Two weeks later I had a phone call - "Is that John Starkey?" enquired a voice; "This is Ed Niles. Your car is for sale in the LA Times." I had to sell almost everything I had and go on my knees to my bank manager (seriously! - but he is a kind man) but I bought 0911 GT and never regretted that decision. Together, we three times competed in the Mille Miglia re-run (in company with my long-suffering wife!) and, during twelve years of ownership, steadily rebuilt the car mechanically.

Although the car left me some years ago for a new life which would start with a ground-up total restoration, my enthusiasm for the car still remains; if some of that enthusiasm spills over in the following pages, I shall be well content.

John Starkey

250 GT Berlinetta

1

THE DEVELOPMENT HISTORY

Ferrari, as a company, has been building chassis destined to be clothed with closed coupé or Berlinetta bodywork almost since Enzo Ferrari himself first set up his own car making business in Maranello, just after the second World War.

Several 166s and 195s were bodied by Touring and Vignale in the late 'forties and early 'fifties with tight little two-seater coupé bodywork, known as "*Berlinetta*" in Italy, literally meaning: "little saloon."

a class of competition, that of "*GT*" (*Gran Turismo*) cars, which was very popular in Europe then. This class demanded closed bodywork with two seats; in Italian, a "*Berlinetta*" as opposed to a "*Berlina*" (a four-seater saloon/sedan). It also became apparent to these owners that a Berlinetta, properly designed with regard to airflow, would usually achieve a higher top speed than the equivalent open car.

Bearing in mind the nature of the great road races, with their long

An early Ferrari Berlinetta: a Touring-bodied 195 car. (Author's collection)

Some Ferrari customers simply wished for the comfort of a closed car, whilst others ordered theirs in the full knowledge that they would be entering

straights (Mille Miglia and Carrera PanAmerica, for example), which took place in this period, closed cars, with their streamlining and better protec-

tion from the elements, became quite popular, even though racing cars were traditionally open.

After the tragic Le Mans race of 1955, in which over eighty people were killed when Pierre Levegh's Mercedes 300SLR catapulted into the crowd, many countries looked to slow down racing cars (Switzerland banned racing outright) and so the GT class, particularly for cars with a cylinder capacity of up to three litres, became popular, culminating in the World Championship for Manufacturers being contested by this type of car in the early 1960s and giving Ferrari World Championship victories with his fabulous GTO.

The direct precursor of the 250GT Ferrari Tour de France Berlinetta was the 250MM series, named after the Mille Miglia race which Giovani Bracco, the Italian hillclimb champion, won with a Ferrari in 1952.

The car Bracco drove was a Vignale-bodied 250S Berlinetta, chassis number 0156 ET, very similar to the then current 225S of 2.7 litres. The car's structure comprised a simple chassis of two tubular side members held together by tubular crossmembers and a network of lightweight tubular outriggers to which the lightweight bodywork was fastened. Suspension comprised unequal length wishbones and a transverse leaf spring at the front, Houdaille lever dampers all round and semi-elliptic rear springs, shackled at both ends and with two radius arms per side. Brakes were huge, steel-lined aluminium drums which filled the inside of the Borrani wire wheels.

The engine used in 0156 ET was the classic Colombo V12 with the cylinder bores opened out to 73mm to accompany the usual stroke of 58.8mm. This brought the capacity of the engine up to 2953.211cc: suitable for the class of GT cars of up to three litres. Thus was adopted the classic bore and stroke of the 250 series which did not alter during the series lifetime: 73 x 58.8mm. An output of 230bhp at 7500rpm, as seen on the factory dynamomoter, was claimed for chassis number 0156 ET.

Three 36DCF/3 Weber carburettors were fitted on the first 250MMs; later examples sported three magnificent 36ICF/4C (4-choke) instruments.

The typically neat installation of the Colombo V12, here fitted with the beautiful 4-choke Weber carburettors typical of those installed on the 250MM engine.
(Author's collection)

The wheelbase was 51.1 inches (1300mm), whilst rear track was 52.0 inches (1320mm). For the Mille Miglia, Bracco's car was fitted with a five-speed non-synchronized gearbox but, later on, a four-speed all-syncromesh gearbox using Porsche-type baulk rings was fitted to 250GTs.

Legend has it that Bracco incessantly swigged brandy and smoked cigarettes passed to him by his co-driver, Rolfo, during the race. What is certain is that the car's tyres were finished by the time he reached the Ravenna services and he was forced to use some undersize ones as there were none of the proper size available. Bracco easily overhauled the works-entered Mercedes 300SLs to win, and came home to a rapturous welcome.

In 1952 the Bracco car was also

0156 ET, the 250S - which Bracco drove to win the 1952 Mille Miglia - as imported into America with a different nose. The car has now been restored to original configuration.
(Author's collection)

entered in the classic 24 Heures du Mans with Ascari and Farina driving, but the early pace was so hot that the Ferrari retired. Franco Cornacchia, one of Ferrari's most active dealers, had a 3-litre Colombo-designed V12 installed in his 212 Inter chassis, 0237 EU, and another 3 litre unit was used in Roberto Rosselini's car - 0265 EU.

Ferrari produced thirty-two cop-

ies of Bracco's winning 250S, naming them "Mille Miglia", or "MM" for short, to commemorate the great victory. The open bodies (Spiders) were built by Vignale and the Berlinetta bodies by Pininfarina, although two Berlinetta bodies were built by Vignale. Thankfully, the Mille Miglia winning car has survived to the present day; at one time the nose was repaired with a

The magnificent 4-choke Weber carburettors of a 250 Mille Miglia. (Author's collection)

A 250MM Pininfarina Berlinetta in 1984 taking part in the Mille Miglia retrospective. (Author's collection)

slightly different shape to the original design but, today, 0156 ET is right back to how she looked when she first emerged from the factory. 0334MM, almost the twin of 0156 GT, has broadly similar lines to 0156 ET, but has more chrome decoration and more room for luggage.

It is in these beautiful little MM coupés that one can see the genesis of the later "Tour de France" Berlinettas.

They combined lightness (2080lb/943.48kg), streamlined bodies and excellent handling (particularly on tight circuits) with a very powerful and reliable engine, all of which applied to the 250GT Ferrari Berlinetta in its various forms, culminating in the GTO.

By 1954, Ferrari had built the last of the 250MMs and the company then introduced a completely new chassis with a wheelbase of 102.3 inches

(2600mm). Ferrari's designers discarded the transverse leaf spring front suspension, substituting coil springs and an anti-roll bar to improve road-holding. At the rear, the oval chassis tubes now swept up and over the live rear axle.

To this new chassis was fitted the Colombo-designed *Tipo* 112 "small-block" V12 engine (as opposed to Ferrari's Lampredi-designed "big-block" engine of up to 4.9 litres) and, thus, the prototype Pininfarina com-

0393 GT. Although built in 1955, this shape (and that of 0425 GT), clearly foreshadows the body shape of the 1957 first version 14-louvre Berlinettas. (Courtesy Pininfarina)

in aluminium, which was retained by the factory from 1954 until 1956 as a development car and occasionally used in competition.

These first Berlinettas were followed by 0393 GT which was the true prototype of the 1957 Competition Berlinetta "Tour de France" body design and had fourteen louvres - de-

petition Berlinetta 0369 GT was born. It was followed by 0383 GT and 0385 GT, which had detail differences in the bodywork such as air outlets just behind the front wheelarches and wind-up, rather than sliding, windows. 0383 GT and 0385 GT were much more similar to the preceding Mille Miglia cars than the next car, 0393 GT. All three, however, had quarterlights fitted. 0357 GT, 0369 GT, 0383 GT and 0385 GT all featured the old Mille Miglia-type transverse leaf front suspension.

0369 GT was preceded by 0357 GT, the first Pininfarina coupé, bodied

0425 GT photographed at the Pininfarina carrozzerie. (Courtesy Pininfarina)

A sad sight! 0425 GT as found by the author in 1983 in southern California. The car has now been completely restored. (Author's collection)

creasing in size towards the rear - running down the panel behind the side windows, and also had headlights behind plexiglass covers. This car was built for Andre Dubonnet, he of the famous aperitif family and a former Bugatti driver of note. Dubonnet entered it for the 1955 Le Mans 24 hour race but, for some reason, it did not make the start. One of the bodywork features which was not to be repeated included a centrally-positioned fuel filler cap on the panel beneath the rear window. This car has now been rebuilt in America.

Incidentally, it should be pointed out that, as these cars were named after events they had won, the "Tour de France" Berlinetta Ferrari only became thus known after De Portago's victory in the 1956 event in 0557 GT. All the preceding cars were simply known as "*Competizione Berlinetta* " by the factory.

The next car to be produced was 0403 GT, which had large fins running from the back of the passenger compartment to the rear of the car in the manner of the 375MM coupé chassis number 0456 AM, the so-called "Ingrid Bergmann" car which was, in fact, bought by Roberto Rossellini. This car was fitted with sliding side windows whilst both 0393 GT and 0403 GT featured a chrome embellisher above the eight ventilation slots behind the front wheel; the frontal aspects of both cars being remarkably similar.

Pininfarina designed and built the

bodies for all these cars, including 0415 GT which reverted to the shape of 0369 GT but with wind-up instead of sliding windows.

A new chapter in the Ferrari story started with the introduction of Scaglietti as a coachbuilder (as well as the builder of Ferrari's chassis). He took the basic elements of the Pininfarina design, but softened them slightly. Naturally, for an Italian coachbuilder, his experiments were all carried out "in the metal" and his first body on the 102.3 inches wheelbase chassis was 0425 GT which was similar to the Dubonnet car (0393 GT) but did not have as much chrome decoration.

With the benefit of hindsight it

may be true to say that Ferrari in the 'fifties and early 'sixties was interested primarily in producing powerful engines with chassis and braking performance being of secondary importance. Certainly, by 1958 the Berlinetta's huge alloy brake drums were outmoded, a shortcoming em-

phasised by the power of the engine.

In 1959 the short wheelbase car would be produced, fitted as standard with Dunlop disc brakes. However, it's probable that a long wheelbase Berlinetta fitted with the same disc brakes would have been as fast around a circuit as the short wheelbase car.

Engine 1955-1959

This V12 unit, designed by Giachino Colombo and first seen in 1947 in 1500cc form, was the heart of the car. Over the years the engine was progressively enlarged by the simple expedient of increasing the bore until it reached its classical 73 x 58.8mm bore and stroke dimensions, giving a swept volume of 2953.211cc.

The Ferrari factory has always had its own foundry, and it used this facility to make all its own alloy castings, thus allowing its designers tremendous flexibility and speed in translating drawings to metal. The cylinder block, heads, sump and timing chain cover were all made from a silicon/aluminium combination known as Siluminum. The block housed the twelve wet cylinder liners and had passages for oil and water and provision for mounting the crankshaft in seven main bearings, which were supplied by Vandervell with whom Ferrari had

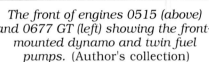
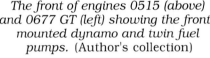

The front of engines 0515 (above) and 0677 GT (left) showing the front-mounted dynamo and twin fuel pumps. (Author's collection)

had a long association (Guy Vandervell bought the first "Thinwall special" from the Ferrari factory in the early '50s).

There were four engine mounting points on the block which mated to the chassis with the minimum of rubber in between. The starter motor was situated on the right-hand side.

The finned and baffled (to prevent oil surge) alloy sump carried the dip-

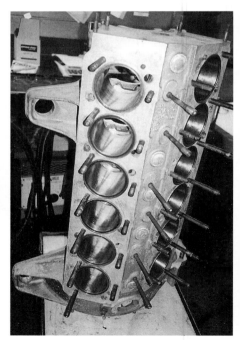

Crankcase of the 128B engine from 0707 GT after honing of cylinders. (Courtesy Lance Hill)

16

Crankcase of 0911 GT showing the staggered bores of a 128C engine. (Author's collection)

The crankcase of 0707 GT, a 128B V12, with the crankshaft fitted. (Courtesy Lance Hill)

The sump of 0707 GT showing both fixed and removable baffles. (Courtesy Lance Hill)

Above: Front timing case with water pump, fan and water pipes shown and, below, pistons & conrods of 0707 GT (Courtesy Lance Hill)

stick and two breather/oil filler pipes, one per side, each surmounted by a polished alloy filler cap containing a gauze element.

In the timing case a three-row chain drove the camshafts, oil, water and fuel pumps and also the Marelli dynamo, upon which was mounted a pulley from which a small belt drove a four- or six-bladed aluminium fan to aid engine cooling.

A fifteen litre copper radiator was used which had a radiator blind, controlled from the cockpit, to aid warming-up in cold weather. The thermostat was in line with the top hose.

The pistons, in alloy with four rings, were made by Mondial initially but, in 1958, a change was made to Borgo pistons with three rings. Piston weight was around 226 grams each but this varied from car to car: favoured customers received the lightest possible set.

The conrods were forged and the big ends split at an angle to allow the rods' withdrawal through the bores with the crankshaft *in situ*: this feature remained until the advent of the

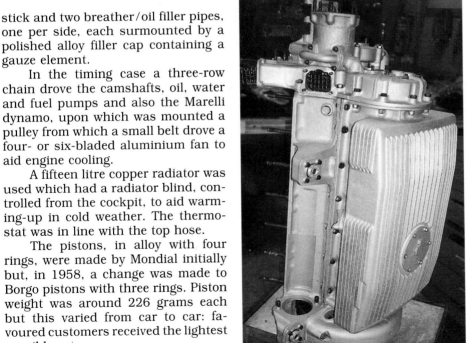

The block of 0707 GT showing the well-ribbed sump. (Courtesy Lance Hill)

128F engine - with its outside plugs and four studs per cylinder - which was introduced in 1959. In this version of the V12 the conrods, which weighed between 435 and 475 grams each, were split across the big ends horizontally, as per normal practice.

The crankshaft was machined from a solid billet of steel with the flywheel and starter ring bolted to the rear. Cylinder liners on these inside-plug engines were press fitted, copper rings, together with a periphery gasket, forming the gas seal. Three cylinder head studs per cylinder were used on these engines, which had factory

The engine compartment of 0911 GT showing the cold air box with lubrication chart attached at the rear. (Author's collection)

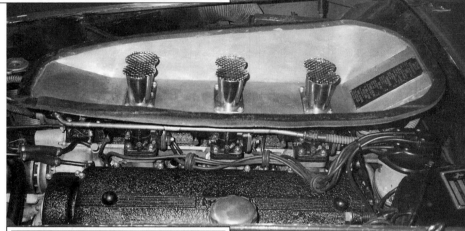

designations of "112" and "128B," "128C" and "128D."

Even in hard racing it was unusual for these engines to fail. 7000rpm was the supposed maximum but 8000rpm was often seen when racing hard. Today, with the stronger metals available, some GTOs being used in historic racing events have left their tachometer tell-tales at 9000rpm and one 1961 Comp. SWB engine is currently producing a prodigious 440bhp at 9300rpm.

The early three stud per cylinder engines had detachable heads in alloy with oil and water passages cast in. One sparkplug per cylinder was used on the inside of the 60 degree vee. Hairpin-type springs (known as "mousetrap" springs in the USA) closed the valves which were set at an angle of 45 degrees. Valve seats and guides were in bronze and the camshafts ran in bearings machined directly in the head. These camshafts were initially of 9mm lift and then of 10mm lift commencing with chassis numbers 0677 GT and 0707 GT: road car camshafts remained at 9mm lift.

The valve timing of the Berlinettas varied between engines as experimentation took place at the factory in search of greater horsepower. Engines built between 1956 and 1958 could deliver 230-250bhp at 7000rpm, whilst engines built between 1958 and 1959 would deliver 235-260bhp.

Three 36DCL/DCZ twin-choke carburettors (40 DCZ6 / 40DCZ3 on the

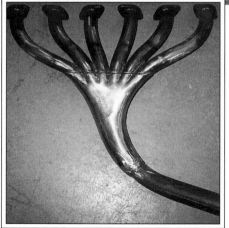

Exhaust manifold of a 128B engine. (Courtesy Lance Hill)

two cars entered for the Le Mans 24 hour race in 1959) were fitted on manifolds which led to siamesed intake ports in the heads. The early (1955-56)

Berlinettas had warm-air induction with no bonnet intake but, from 1957, cold air induction was fitted which featured a bonnet intake: a large alloy dish with a rubber seal which clamped to the underside of the bonnet.

Fuel pressure for starting was supplied by an electrically-operated Marelli fuel pump situated at the rear of the car and controlled from a dashboard switch. This pump could also be used to supply extra fuel at high speed or in very hot weather conditions. One, sometimes two, mechanical pumps were situated on the engine to supply the fuel once the electric pump had primed the system. The starter motor was fitted on the right side of the cylinder block, located low down underneath the exhaust manifold on that side.

On the left, the original fuel tank from 0707 GT with the replacement tank on the right. (Author's collection)

The front suspension of 0515 GT whilst under rebuild. (Courtesy Ed Niles)

Houdaille dampers disassembled for remachining. (Courtesy Lance Hill)

Competition car early inside plug engines came with twin Marelli distributors, coils and condensors, while "street" 250GT coupé engines tended to have only one distributor.

Each exhaust manifold comprised six pipes at the head end leading into the single outlet to the exhaust system, which was fitted with one silencer per bank; the exhaust finally exiting through four chrome-tipped pipes at the rear. The radiator held 3.3 gallons (15 litres) of coolant and fuel tank capacity was 26.4 gallons (120 litres) in the 1954-55 cars, while later cars had a larger tank capacity.

The fuel tank was filled through a huge filler neck surmounted by a "Monza" alloy fuel cap, and was situated inside the trunk on early cars and fitted flush with the bodywork on later cars. The sheer size and capacity of these tanks has startled many a fuel pump attendant! The largest tank fitted was probably that which equipped 0707 GT, which held 30 gallons (137 litres). A mechanism to lock the cap from the inside of the boot (trunk) was fitted to some cars.

Chassis 1955-1959

Looking at the chassis of these "Tour de France" Berlinettas today, the first thing that strikes one is their simplicity. Indeed, in an age when Ferraris are looked upon with awe by the majority of the population, it's good to know that, provided these cars are well maintained, they are really very reliable. My old car, 0911 GT, competed in no less than three Mille Miglia re-runs, besides doing the real event twice in

the 'fifties, and was driven to Italy and back each time with no real problems.

Two long tubes, interconnected by five crossmembers and with X-bracing in the middle, make up the basic chassis frame. The drawing office at the factory designated this frame internally as the "Tipo 508." Through the following five years the suffixes "B," "C," "D" and "G" would denote slight variations of the chassis in the form of strenghening tubes and the location of dampers and springs.

The wheelbase remained at 2600mm (102.3 inches) throughout production, except for chassis numbers 0403, 0415 and 0425 GT which each had a chassis with a wheelbase

The chassis of 0677 GT whilst under rebuild.

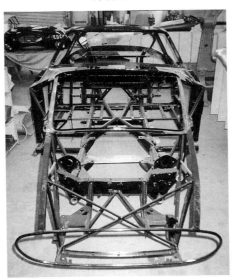

reduced to 2480mm (97.63 inches). The chassis were supplied to Ferrari by Vaccari of Modena.

On all the Berlinettas the front track was 1354mm (53.3 inches) while the rear track was 1349mm (53.11 inches). The rear suspension had the usual semi-elliptic springs with two tubular trailing arms on each side but the front suspension now used two coil springs instead of the single transverse leaf previously employed. An anti-roll bar was also fitted at the front and lever-type Houdaille dampers were in-

The rear competition drum brake of 0515 GT. (Courtesy Ed Niles)

The huge drum brakes: they completely filled the inside of the 16-inch Borrani RW3264 wheels. (Author's collection)

stalled all round. Incidentally, Phil Hill once remarked to the author that these dampers soon lost their effectiveness when the cars were raced hard. Undoubtedly, this led to the adoption by Ferrari of more effective telescopic dampers in the later short wheelbase Berlinetta.

The long wheelbase Berlinetta's ZF steering box had ratios from 1:17 to 1:20 and the steering column was adjustable for length. The propellor shaft had a rubber universal joint connection to the differential made by "Saga" and a Cardan joint made by "Fabbri" connected to the gearbox.

The clutch and brake pedals were contained within an aluminium casting bolted to the chassis and the pedals were hinged at chassis level, whilst the accelerator linkage worked through an intricate series of bellcranks and rods.

Up to 1957, a multi-disc clutch made by Frendo was fitted but, later, a single-plate clutch by Fichtel and Sachs was used which greatly benefited the ease with which the car could be moved away from a standstill.

The gearbox for 1954 and 1955 cars was from a type 342 America fitted with Ferrari syncromesh; later cars benefited from a new four-speed gearbox fitted with Porsche syncromesh and featuring evenly spaced close ratios. The gearbox casing was in alloy (some were ribbed for strengh and heat dissipation) whilst the gearbox also had its own oil pump. A central or offset gearchange could be ordered.

The rear axle had a ZF limited-slip differential with an alloy case. Five differential ratios were offered ranging from a 7 x 32 low up to a 9 x 35 high. With the latter ratio fitted, the car was capable of 66mph in first gear at 7000rpm whilst top speed at these engine revolutions was given as 156.6mph. The 7 x 32 ratio provided 53.4mph in first gear and top speed was 125.5mph. That the gearbox ratios were delightfully spaced can be gauged from the fact that second gear with the lowest axle ratio gave 75.8mph with 101.9mph available in third gear.

Huge finned aluminium drum brakes of 360mm (14.17 inches) diameter were fitted to all four wheels, some with six cast-in ventilation slots, as seen on the earlier Grand Prix cars. Brake shoes were of aluminium also, with competition (hard) linings riveted on, whilst the drums themselves were lined in steel. One, sometimes two, master cylinders were fitted and a small booster was attached to the front of the chassis. No servo was fitted by the factory on these cars. Borrani wire wheels were always used; RW3077 500 or 550 x 16 from 1954-55 and 600 x 16 RW3264 for the rest of the series. The two-ear knock-on spinners were fitted

Electrical panel and fuses of 0707 GT during restoration. (Courtesy Lance Hill)

The seats of 0707 GT under restoration. (Courtesy Lance Hill)

although a number of cars today feature the three-eared variety. A three, sometimes four-piece undershield was fitted to the chassis with holes to accommodate inspection and oil change to engine, gearbox and rear axle.

All trim was in aluminium and the windscreen was in toughened glass by "Sekurit" with side and rear window made of perspex to cut down weight. A comprehensive toolkit, which included a wheel hub puller as well as spanners and a jack and copper hammer, was supplied.

Today, one often sees rubber strip fitted to suppress draughts, but this has been owner-fitted retrospectively. All long wheelbase Berlinettas are left-hand drive and serial numbers range from 0357 GT to 1523 GT. A total of 94 long wheelbase competition Berlinettas were produced by the factory.

250 GT Berlinetta

2

CONTEMPORARIES OF THE "TOUR DE FRANCE" BERLINETTA

Earlier in the book, I mentioned the 250GT Europa roadgoing coupé and it's worth digressing from the competition Berlinettas to take a look at these essentially roadgoing cars. Nearly all of Ferrari's road cars benefited from experience gained by the racing versions, particularly with regard to engines, brakes and suspension.

250GT Europa

The Europa came in two forms. The first, built from 1953-54 and introduced at the 1953 Paris Salon, featured a linered down big block engine of the type designed by Lampredi and used in 4.5 litre form in Ferrari's 375 America and Mille Miglia cars. The Europa's ladder-type chassis, gearbox and brakes were identical to those of the 375 model. A total of twenty-one of these cars were built and they were noted for their truck-like, heavy steering. The model was replaced in October 1954, at the Paris Automobile Show, by a new "GT" version.

The new GT featured the Colombo-designed small block engine with a bore and stroke of 73 x 58.8mm and

The 250GT Europa.

0373 GT today. (Author's collection)

The interior of 0373 GT today. (Author's collection)

Michelotti and Vignale for Princess Liliane de Rethy of Belgium. It was to be the last Ferrari bodied by Vignale, Pininfarina becoming the main coachbuilder for Ferrari road cars with Scaglietti executing most of the competition bodywork.

These first examples were the true forerunners of the outstanding 250GT road cars and were noted for their good roadholding and light steering as evidenced by 0357 GT, the original Paris show car, which came third in the 1956 Tour de France when driven by Olivier Gendebien and Michel Ringoir. Incidentally, Gendebien had expected

Olivier Gendebien taking part in the 1956 Liége-Rome-Liége rally of 1956. Gendebien, accompanied by Pierre Stasse, came in 3rd. The car is 0373 GT, a 250GT Pininfarina-bodied Europa, owned at the time by Philippe Wascher, Gendebien's cousin. (Courtesy A. Callier)

designated "*Tipo 112*" by the factory. In effect, this was a touring version of the engine used to power the 250MM, the main difference being that two-choke (instead of four-choke) 36DCF Weber carburettors were used, the manifold having siamesed inlet ports of the type employed in the early Tour de France cars. Chassis length was reduced to 2600mm (102.3 inches), in effect the same as the competition cars, whilst the coil spring type (as opposed to the transverse leaf spring of the first-series Europa), front suspension was used. The four-speed gearbox and multi-plate dry clutch were the same type as those used in the 342 America.

Thirty-six Europa GTs were built up to the beginning of 1956, commencing with chassis number 0357 GT. Most were bodied by Pininfarina and one, 0359 GT, was bodied by

The 1956 Geneva Motor Show 250GT prototype with body by Pininfarina.
The series was later built by Boano.
(Courtesy Pininfarina. Kurt Miska collection)

to drive 0503 GT, a works entered 250GT Berlinetta, in this event but, at the last moment, it was allocated to Alfonso de Portago who went on to win the event outright.

0373 GT, another Pininfarina bodied 1954 coupé bought by Philippe Wascher, Gendebien's cousin, from Jacques Swaters' Ecurie Francorchamps, was used by Gendebien and his navigator, Pierre Stasse, to come third in the 1956 Liége-Rome-Liége rally which was won outright by Willy Mairesse in a Mercedes 300SL. The engine of 0373 GT delivered a claimed 215bhp, whilst the car sported large wire mesh stone guards to protect headlights and spotlights. 0373 GT's weight, incidentally, was stated as 1100kg/2425lb, considerably more than a Berlinetta's.

Boano series 1955-1957

The next series of 250GT coupés was built on the same chassis as the Europa GT but with a Pininfarina body design executed by Carozzeria Boano situated at Grugliasco, near Turin. It's noticeable how the road car design was constantly developed, both in engine and chassis, by reference to the competition cars. In the Boano series such development manifested itself with the appearance of the "*Tipo* 128B" and "*Tipo* 128C" engines. Chassis numbers 0403, 0405, 0407 and 0409 GT featured 128 engines and they

A Boano Coupé being used in competition sometime in the 1950s/1960s.
(Courtesy Kurt Miska)

The classic lines of 1081 GT, a 1958 Pininfarina coupé.
(Author's collection)

precede 0503 GT, the first Berlinetta to be fitted with this variant of the V12. Of course, in the touring cars these engines were slightly detuned by employing a single twelve lead distributor and a lower compression ratio.

Otherwise a stronger crankshaft was new to the *Tipo* 128 series of engines, together with reinforced conrods and main journals. A strange gearchange pattern appeared with these cars with first and second gear on the right and third and fourth on the left-hand side. Reverse was on the far left, parallel with first and third gear.

The Boano/Pininfarina body was a very clean design executed mainly in steel; some cars were an exception to this rule by being panelled in aluminium, presumably for owners who wanted to take part in mild competition, but with the benefit of interior trim and better weather proofing. 0555 GT was just such a car and took part in events such as the Mille Miglia: it was bodied by Pininfarina. Bonnets and boots were panelled in aluminium on most Boano-built cars. About ninety cars were built in Grugliasco, and were followed by a further series of between forty and fifty cars with a slightly raised roof line.

The bodies of these Boano cars were built at the same premises as before, but the company was now known as the *Carozzeria* Ellena. This was because Boano had left to take charge of FIAT stying and so Boano's partner, Lucianno Pollo, continued the business, joined by his son-in-law, Ezio Ellena, from whom the company took its new name.

With these Boano body shapes, Ferrari had now moved on to a form of "series production" as mentioned in the brochure for 1956 and, from this point, fewer and fewer special designs for the road or show were produced. With the next design this process went even further as, in 1957, production of the 250GT bodywork reverted to Pininfarina.

Pininfarina series 1958-1960

A noticeable difference in this next design was the lowering of the waistline, allowing greater window area, whilst still using the wrap-around type rear window of the Ellena/Boano coupé. These were most elegant bodies and, even today, one can still admire the gracefulness of this two-door, two-seat coupé. Unlike contemporay American designs, Pininfarina eschewed ornamentation for its own sake and this resulted in a timeless shape. The car was introduced at the Milan Auto Show on 25th June 1958; at the same time the coachbuilder's name was changed from the two separate words of Pinin Farina to one: "Pininfarina."

By now, disc brakes were in production use on both the Jaguar XK150 and the Aston Martin DB4, the latter car being bodied by Touring of Italy. It was in 1958 that Mike Hawthorn had disc brakes mounted on his Formula One Dino 246 racing car with which he went on to win the Formula One World Championship. However, the change to disc brakes for Ferrari road cars would not happen for another year as Enzo Ferrari still believed his drums to be at least as good: the introduction of discs on the 1959 short wheelbase Berlinetta would convince him otherwise.

Around three hundred and fifty examples of the new Pininfarina bodystyle were built, making this far and away the most successful venture into road car sales Ferrari had ever experienced. The later 250GT coupés featured 128F engines with sparkpugs on the outside of the vee and coil, instead of hairpin valve springs as per the Competition Berlinettas, as well as disc brakes. Overdrive was also featured on some of the later cars. With five rear axle ratios giving from 124mph to a theoretical 155mph, and space for two people plus an "occasional" seat, a parcel shelf and spacious boot, these cars were a great success for Ferrari.

I was fortunate enough to own the London Motor Show car of 1958, 1081 GT, and, even though the car was then twenty-one years old, its engine, steering and suspension still impressed. I had to sell it to make way for my own "Tour de France" Berlinetta, chassis number 0911 GT, but still remember with affection its smooth and rapid progress and timeless lines.

When viewing the history of the Ferrari marque, it's always best to bear in mind that Enzo Ferrari was only really interested in building racing cars and seeing them victorious. He only built road cars to finance the competition cars and, frankly, the fact that road cars did not interest him can be seen in his basically "couldn't care less" attitude towards his road car

The 250GT Spyder California.
(Author's collection)

customers. It was this attitude which was to spur a certain Ferrucio Lamborghini to build his own road cars to compete with Ferrari.

Cabriolets & California Spyders

Apart from the coupés made during this period (1954-59), Ferrari also produced two quite different open cars: the first version of the Cabriolet and the California Spyder. These were aimed, as were the Berlinettas and the Coupés, at two different markets.

The Cabriolets were the open equivalent of the coupés with their tamer engine timing, creature comforts and softer ride, whilst the California Spyders were essentially open versions of the Berlinetta.

We'll deal with the Cabriolets first. With the exception of the first one produced, 0461 GT which was bodied by Boano, all were steel bodied by Pininfarina and most possessed his typical design features of the period - covered headlights and kicked-up rear wing line behind the doors stretched out to almost fin shape, with the rear lights set into the edges. Chassis number 0655 GT was used by Peter Collins as his personal transport whilst he was a member of the Scuderia Ferrari and this car featured as an elbow rest an unusual cutaway in the top of the driver's door. On the Cabriolets, the top would fold down to be stowed behind the seats. Some of the cars also featured air vents behind the front wheels, (although most did not), whilst the chassis and engine were identical to the then current coupés.

The Series 2 Cabriolets were introduced in 1959, and were very similar in design to that year's coupé, being more conservative than the earlier series. As these fall outside the time scale of this book, I will simply recommend to the reader Antoine Prunet's superb book on the subject of Ferrari's road cars.

The Spyder California referred to earlier, although designed by Pininfarina was, like the Berlinetta, actually built by Scaglietti of Modena. The California was the result of Luigi Chinetti, the Ferrari distributor for Noth America, realizing that an open version of the Berlinetta would sell very well in the warmer temperatures of the USA. Unlike the Berlinettas, however, nearly all Californias were bodied in steel with the doors, bonnets and boots made in aluminium. All the long wheelbase Californias were completed by March 1960. A very few cars were built with all-alloy bodies and "Monza" type filler caps leading to a larger fuel tank. These cars were nearly all meant for competition and some had a windshield approximately one inch (25.4mm) lower than their road-going cousins.

In 1959, the first year the California was raced in international competition, one car driven by Ginther and Hively won the GT class at Sebring and came in nineth overall whilst, at Le Mans that year, a California (1451 GT) driven by Bob Grossman and Tavano

was placed fifth overall. Bob Grossman continued to campaign this car in US SCCA events with good results until he sold it: the car has now been perfectly restored.

Another California was fitted with an engine in *Testa Rossa* (Red Head) tune and was entered by the *Scuderia Serenissima* for the Sebring 12 hours in 1960 where it finished in eighth place overall.

Doing away with the roof of a long wheelbase Berlinetta meant that a flat rear deck was fitted behind the cockpit whilst the windscreen was sharply raked. Apart from these changes the California and the Berlinetta were very similar. The array of instruments and stark interiors were as the Berlinetta and so were the states of engine tune, although few were delivered with the ten millimetre lift camshafts. Most engines featured three thirty-six millimetre Weber carburettors and twin distributors; some of the later cars were fitted with disc brakes.

Forty long wheelbase Californias were built, followed by a longer series of short wheelbase cars. As I've pointed out, these cars are outside the scope of this book but, as further reading, I can thoroughly recommend *The Spyder California* by George Carrick.

These, then, were the cars which Ferrari stalwarts could buy if they did not wish to go in for the hurly burly of GT racing in the 1950s. However, If the customer was looking for a GT car to

The Mercedes-Benz 300SL of 1955.
(Author's collection)

Interior of the Mercedes-Benz 300SL. (Author's collection)

compete against Ferrari with success, he'd have a very hard - if not impossible - job to find another car with all the virtues of the 250GT Competizione Berlinetta.

Lancia Flaminia

Lancia made the Flaminia with a V6 engine of 2600cc which, with lightweight Zagato "Double Bubble" bodywork, could always be counted upon to give the Ferraris a run for their money, particularly in Italian events where sheer weight of numbers helped. However, that extra 400cc and the continuous development of Ferrari nearly always told against them.

Mercedes 300SL

In 1955-56 the GT car to enter and win with was the Mercedes 300SL: a formidable contender. Lightweight versions of this fuel-injected straight six, gullwing doord car won most GT events up to 1955. Then the 250GT Berlinetta appeared and the 300SL was beaten; even Mercedes-mounted Stirling Moss had to settle for second place on the 1956 "Tour" behind De Portago and Nelson in their competition Berlinetta, chassis number 0557 GT.

Jaguar XK/Aston Martin DB2/4, DB3 & DB4GT

If you were British, naturally you would want to buy a car that was "made in England" but there were only two real contenders which had been built in sufficient numbers to qualify: the Jaguar XK and the Aston Martin DB2/4 or DB3 or, much later in 1959, the DB4GT.

The shapely rear of the Mercedes-Benz 300SL. (Author's collection)

The Jaguar XK of 1956 to 1957 which could have competed against the 250GT, was the fixed head coupé XK140, basically a rehash of the XK120, a design already nine years old. The XK140 did once show just what it was capable of when Bolton and Walkinshaw entered the 1957 Le Mans with a fixed head coupé.

With some factory assistance, such as a high performance cylinder head and a long-range fuel tank, the XK surprised everyone by getting well into the top ten before a dispute over oil replenishment saw its withdrawal very near the end of the race. Just to rub

27

The 1959 Aston Martin DB4GT with "touring" bodywork. (Author's collection)

Below: A Jaguar XK140 fhc modified for racing in much the same way as the Bolton/Walkinshaw Le Mans car. (Author's collection)

salt into the drivers' wounds, it was later discovered that the organizers were at fault in disqualifying them. Ian Appleyard, the great XK120 exponent in international rallies, used an XK140 fhc, registered VUB 140, to take second place in the RAC Rally and Guyot used an XK140 to win his class in the Mille Miglia.

Sadly, these were the XK140's only real entries in serious international racing; none of them were ever entered for the Tour de France. However, it's unlikely the Jags would have done well in the "Tour" as it placed a premium on good handling, good brakes and nimbleness; all of which the Ferrari possessed by virtue of its lightness and big brakes. The Jaguar had to make do with narrow iron drum brakes until

the advent of the XK150, with its Dunlop disc brakes. But this XK, fatter now as well as being even more directed at the roadgoing market was hardly campaigned at all in Europe.

The roadgoing Jaguar of this period which was used in competition was the 3.4 and, later, the 3.8 litre MkI and MkII saloon. In the Tour de France, in particular, they entered the touring

The twin-plug straight-six engine of the Aston Martin DB4GT.
(Author's collection)

class and did well, coming third in the scratch class in 1958 and winning in 1959. If only Jaguar had homologated a hardtop D-Type!

Aston Martin had a few entries in continental events with the DB3, an update of the DB2/4, but it could never match the Ferrari Berlinettas, being heavier and having a smaller engine than the 250GT.

In 1959 Aston Martin produced the DB4GT, the first serious attempt at matching Ferrari's game. With triple Weber carburettors, 3.7 litres and twin ignition, it was a worthy contender, particularly in Zagato lightweight form, but, by then, Ferrari had moved on to the short wheelbase car and the Aston never beat it.

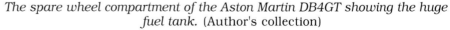

The spare wheel compartment of the Aston Martin DB4GT showing the huge fuel tank. (Author's collection)

A 1956 Berlinetta at an American race. (Author's collection)

Apart from the examples already noted (0357GT, 0369 GT, 0383 GT, 0385 GT, 0393 GT, 0403 GT, 0415 GT) which were bodied by Pininfarina, and with the exception of a few cars with bodies by Zagato, all the 1956 cars were bodied by Scaglietti whose bodyshop was in Modena (today, Scaglietti still builds the bodywork fitted to all Ferraris).

All the bodies intended for competition Berlinettas were handmade in aluminium with small diameter steel tubes attached to the chassis acting as a framework. Windscreens were of "Sekurit" shatterproof glass and side and rear windows were of perspex.

In 1956 the car's shape was obviously developed from the 250MM Pininfarina Berlinetta and the 375MM Berlinetta. From the front the usual Ferrari "egg crate" grille fills the elliptical radiator intake with the nose then sweeping up to the windscreen. These early cars used warm air induction with a small blister on the bonnet but, later, most were altered to a hood scoop giving ram-type induction to a cold air box sealed at the edges, with sorbo rubber and sealing beneath the bonnet (hood) when the bonnet was clamped down by the springloaded and chromed twist-and-drop fasteners and

The 14 louvres of 0425 GT; built in 1955 but foreshadowing the 1957 1st version bodywork. (Author's collection)

leather straps at the rear of the bonnet opening. The headlamps (Marchal) were set low down and covered by plexiglass cowls which followed the contours of the wings (fenders) with sidelights beneath; the wing shape was gently rounded on top.

On the lower sides of the front wings, behind the front wheelarches, were, first, an engine ventilation slat divided into eight segments on each side and then a small ventilation door just in front of the door openings. This door allowed cool air to reach the cockpit, via hoop-reinforced trunking which led from an intake behind the radiator

0509 GT, owned for many years by Bob Dusek in the USA.

grille to the driver's (and, in some cases the passenger's) footwell, when a tee shaped lever, situated beneath the dashboard, was pulled.

Behind the door opening, the rear wings swelled slightly to fit into a curved tail. The small boot lid, operated by a chromed pushbutton set beneath the lid, opened to provide a little luggage space, plus access to the spare wheel and fuel filler neck with alloy quick-release cap.

Set into the black crackle-finish (smooth painted on early cars) alloy dashboard, was a large, wood-rimmed Nardi steering wheel with the prancing horse badge placed in the centre acting as a button to activate the air horns. The steering column extended from a separate binnacle which housed a 300kph speedometer plus a tachometer which read to 8000rpm (red lined at 7000rpm). The oil pressure gauge was situated between these two dials - as befitted its importance - and read in pounds per square inch. Placed on either side of these large dials was a water temperature and an oil temperature gauge. Four green and red lights provided information of the operation of the ignition, fuel pump, heater fan and lights. Across the rest of the dashboard, from left to right, were the fuel gauge, the ignition switch, a clock and a red "T" sign which glowed if the coolant temperature reached 100 degrees centigrade. A glovebox with no lid was in front of the passenger.

Beneath the dashboard was a chain which, when pulled, operated a radiator blind to assist warm-up. Also under the dashboard was a row of

The side vent of 0425 GT, complete with Pininfarina badge.
(Author's collection)

chromed switches and bakelite knobs for operation of the lights, fuel pump, wipers (non self-parking), washers, fog or the long range lights (made by Marchal and set into the grille opening), dashboard lights and a rheostat. A fuseboard was provided beneath the

dashboard on the passenger's side. A drum-like heater and demister was situated in the centre, above the gearbox tunnel, and had two small doors which could be opened to admit hot or cold air into the interior. The handbrake was a pull-and-twist type situated under the dashboard to the right of the driver.

The accelerator pedal was connected to the carburettors via a series of beautifully-fabricated levers and bellcranks. The clutch pedal operated the multi-disc clutch through a system of levers set on splined shafts, whilst the brakes were operated via a hydraulic system, the alloy reservoir being mounted on the engine compartment bulkhead. Some later cars featured a twin circuit braking system.

0415 GT. Correct inside-plug Ferrari V12 now nestles where an American V8 temporarily resided. (Courtesy R. Pas)

0515 GT whilst under restoration.
(Author's collection)

Seats were covered in leather or leatherette to customer choice with a headrest for the passenger; carpeting was fitted on the floor and was also used to cover the space behind the seats. A leatherette material with a distinctive diamond pattern was used beneath the dashboard and sometimes over the gearbox cowling. Leather straps were fitted to restrain luggage in the rear whilst a dipping mirror, (sometimes with a small light installed) and a plastic headlining were fitted. There were no sun visors.

The passenger's door was lockable from the inside, whilst the driver's door was locked from the outside. Sliding side windows were fitted to some cars and the wind-up type to others, according to customer preference. The interior remained basically the same for all the Berlinettas.

Of the 1956 cars, two, 0515 GT and 0537 GT, were bodied by Zagato, the company which has always been famous for producing extra-lightweight bodies for sporting and competition cars for such famous marques as Alfa Romeo, Abarth, Aston Martin and Lancia.

Both of the Zagato-bodied cars featured a characteristic of Zagato's coupé bodies, a "Double Bubble" roof. 0515 GT was heavily campaigned by Camillo Luglio after which it found its way to America; there Ed Niles bought and sold it no less than five times, before finally restoring it to its former glory to take a "best of show" at Pebble Beach concours.

Racing in 1956

It is a fact that the "Tour de France" Berlinettas won more races than either of their illustrious successors, the SWB and the GTO. This is down to two factors; length of service and the number of races that could be entered by the Berlinettas.

Length of service really points to the fact that the long wheelbase car was used in frontline GT competition

The spare wheel compartment of 0515 GT showing the filler cap.
(Courtesy Ed Niles)

0515 GT, the first Zagato body on a long wheelbase Berlinetta, photographed before its rebuild.
(Courtesy Ed Niles)

The interior of 0515 GT in the 1960s.
(Courtesy Ed Niles)

for four years; 1956 to 1959. Add to this the fact that a well-driven car was still capable of taking the occasional GT victory in 1960, and you can see that the cars had a a very long competition life.

There were probably more events in which the cars could compete in the relevant years. The 250GT Berlinetta was equally at home on a rally, a hillclimb or a circuit race - and must be one of the very few cars which could claim that. Some drivers, particularly the Italians, specialised in the long European hillclimbs such as Pontedecimo-Giovi and Aosta-San Bernardo, whilst it was mainly the French and Belgians who entered the "Tour de France".

1956 saw the real start of the 250GT long wheelbase Berlinetta's competition history, although Alfonso de Portago had taken 0415 GT to Nassau for the "Speed Week" event and won the GT race in December 1955. Come 1956 and Gendebien, once again with his cousin, Jacques Wascher, won the GT class in the Tour of Sicily as well as coming in fourth overall in another Scuderia Ferrari works car. In that year's edition of the Mille Miglia, Gendebien was fifth overall (and GT class winner) in 0503 GT, a Scuderia Ferrari-entered car. In this very wet event he left the road four times and spun three times; at one point, the

doors had to be tied closed with string! Although very battered, the Berlinetta still rewarded its crew with a great result.

The 250GT Berlinetta, particularly the long wheelbase car, was nicknamed the "Tour de France" model for the simple reason that it won the event every year it was entered, that is, four times. No less than three of these "Tour" victories were won by that doyen of GT drivers, Olivier Gendebien.

The "Tour" comprised over three thousand road miles around France, sometimes into Italy, Germany and Belgium. It took five to six days and consisted of up to six races, two hillclimbs and a sprint. In 1956 the event was won by Alfonso de Portago with Edmund Nelson as his navigator in 0557 GT, with Gendebien coming in third in the first Pininfarina coupé made, the ex-works development car, chassis number 0357 GT. Stirling Moss was second in a Mercedes 300SL and this event signalled the end of the

300SL as a GT contender: in fact, no 300SL ever won a race in which a 250GT Berlinetta was entered ...

Apart from this, De Portago took 0557 GT to first overall and first in the GT class at Montlhéry in October in the Coupé de Salon and then won the GT

0537 GT, the second Zagato long wheelbase Competition Berlinetta.
(Author's collection)

class in the Rome Grand Prix. The car even boasted a radio!

Italian "Gentlemen" drivers had fine seasons in the 250GT, Edouardo Lualdi winning five events and taking three second places, mainly in hillclimbs in 0539 GT, whilst Camillo Luglio, driving Zagato-bodied car number 0515 GT, had five victories and two second places.

0509 GT, in the hands of Guiliano Giovanardi, took five first places, two thirds and an eighth overall in the Mille Miglia, whilst Pininfarina Berlinetta number 0383 GT, in the hands of Paulo Lena, was third in the Coppa Inter Europa at Monza (an event for GT cars held just before the Italian Grand Prix) and Lena followed this up with a second place in the Coppa San Marino.

FERRARI
FERRARI

250 GT
Berlinetta

4

1957

0677 GT, the works' car, which was driven by Gendebien, as it is today.
(Author's collection)

In 1957 two new body shapes appeared. The first stretched the car slightly to give a longer, lower effect. The headlamps were still set into the front of the wings but now the bonnet had an air intake leading to the carburettors.

The top of the front wings was angular instead of being rounded and the rear wings were very distinctive: a definite evolution of shape was taking place which would reach maturity in the 1958 bodywork.

The engine air vents now comprised six slats instead of eight and, behind the cockpit, the large, wraparound rear window gave way to four-

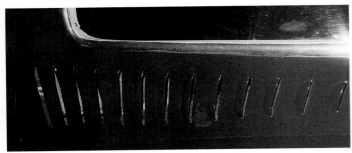

The 14 louvres of 0677 GT. (Author's collection)

The spare wheel compartment of 0677 GT.
(Author's collection)

0579 GT at a Prescott hillclimb event in the 1970s. (Author's collection)

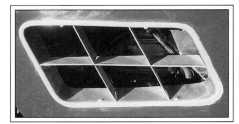

The cooling slots of 0629 GT.
(Author's collection)

The interior of 0677 GT.
(Author's collection)

teen louvres that vented air from the cockpit; this ventilation was controlled by a small door with a bakelite knob behind each seat. Today, these cars are referred to simply as "fourteen louvre cars." Small upright bumperettes with rubber inserts were fitted front and rear.

Chassis numbers 0665 GT and 0689 GT were Zagato bodied.

0707 GT, the last "14-louvre" Berlinetta built, had a special order fuel tank which held 137 litres (30 gallons) and had no fuel gauge; instead having a warning light on the dashboard which glowed when the level became low. This car's assembly sheet added the suffix "20" to chassis, engine, gearbox and back axle numbers. Paul Frere carried out a road test of this car in 1958 and recorded acceleration times of 0-100kph (62.1mph) in 7.6 seconds and 0-160kph (99.4mph) in 15.0 seconds. As well as this, the car did the standing quarter of a mile in 15.0 seconds and the standing kilometre in 26.9 seconds. Frere noted "When reading the acceleration figures, it must be kept in mind that Ferrari clutches would not accept a 'racing start' with spinning wheels and had to be treated with considerable delicacy." I can testify to this, having sat on the startline

Below & top of next page: 0629 GT, a 14-louvre Berlinetta, in California after restoration by Bradley Balles.
(Author's collection)

at Prescott Hillclimb, dropped the clutch at 6000rpm and being rewarded merely with a lot of smoke and an awful smell from the slipping clutch!

On the fourth of April 1957 the next version of the Competizione Berlinetta bodywork appeared and, in this incarnation, the body shape approached maturity. Today, many people seem to think that the 275GTB shape was derived from that of the 250GTO but, with its slots and covered headlights (not to mention rear wing bulges), the similarity between the Tour de France Berlinetta and 275GTB is much more evident.

From the front, the changes could be seen mainly in the headlight positioning; they had moved higher and further back into the wings and were fitted with perspex covers to blend into the wing shape.

The Marchal fog/long range lights now sat outside the grille, either mounted on the thin aluminium bumpers or on brackets situated at the bottom of the grille. Sidelights/direction indicators were recessed into the front wings beneath the headlights.

Moving further back, the air outlet vents from the engine compartment now comprised three slots cut into a detachable panel on each side, whilst

The interior of 0629 GT. (Author's collection)

37

a chrome rubbing strip was fitted immediately below the doors, spanning front and rear wheelarches.

Abaft the doors, the rear wings now bulged into very distinctive forms with the top edges being cut off at their ends and the rear lights inset. Behind the side windows the fourteen louvres growing wheel and tyre widths. At the rear the large Monza filler cap, placed at a corner of the boot lid, protruded slightly from the body surface.

The exhaust system was as before with four chromed tips protruding from beneath the rear aluminium bumper. These bumpers, incidentally, were only fastened to the skin of the bodywork,

1957 2nd version bodywork. (Factory photograph)

had disappeared to be replaced by three slots tapering in size towards the tail; cockpit ventilation was still controlled from the interior by sliding bakelite knobs.

The wheelarches had small lips bulging outward to accommodate not located via heavy steel mountings to the chassis. The Berlinetta's interior was the same as before except for small details such as the courtesy light which was now incorporated into the dipping mirror.

The twin-plate clutch was replaced

38

0707 GT as it arrived in England in the 1960s to be rebuilt. The car had a 1958 front end which was replaced by the correct 1957 1st version nose.

by a Fichtel and Sachs single-plate type which resulted in a markedly smoother take-off from rest and easier driveability: most cars featured a *Tipo* 128B or C engine. The fuel tank now held 35.7 gallons (162 litres).

The engine compartment of 0767 GT as it is today. (Courtesy L. Beltrami)

Racing in 1957

Olivier Gendebien, partnered by Lucian Bianchi, won the Tour de France this year in 0677 GT - a car with which Gendebien enjoyed fabulous success.

Gendebien, this time partnered by Paul Frere, also won the Twelve Heures du Rheims two years running, 1957 and 1958, both times in 0677 GT. It is recorded that an engine in Testa Rossa tune was fitted in 0677 GT for certain important events. It certainly was in use in this event, but in this application the TR engine always used the regular three carburettor set-up instead of the six carburettors employed in the Testa Rossa sports racers.

During 1957 Gendebien also won the GT class in the Mille Miglia, this time with his cousin, Wascher, as his navigator/co-driver. They came in third overall behind two very special Ferrari four cam sports racing cars, one of four litres and one of three point eight li-

tres, known as *Tipos* 335S and 315S respectively. The drivers of the two four cam Ferraris had great difficulty in passing the Belgian driver: he averaged 123.91mph (199.4kph) for the eighty-two mile (131.96km) long stretch from Cremona through Mantua and on to Brescia and won the Coppa Nuvolari for being the fastest car over this last stretch. The engine in 0677 GT had camshaft timing similar to that of a 290MM and it was estimated to deliver some 260bhp. In the 1957 event, six Mercedes 300SLs started the event but none finished, whilst eight other Ferrari 250GTs finished from sixth place downward.

1957 found the Berlinettas in tremendous form and, with more of them being supplied to eager drivers by the factory, the results reflected their superiority. The 250GT's season started with 0629 GT, driven by Lena Palanga, winning the GT class in the Sestrieres Rally held between February 24th and March 1st. Lualdi now had a 1957 Scaglietti first series car and his season included a superb seven first places in the GT class, three overall victories and a second place in the Coppa Inter Europa.

Olivier Gendebien, as already mentioned, had an excellent 1957 sea-

1957 Mille Miglia. Gendebien and Wascher depart Ravenna in 0677 GT. They go on to win the GT Class and come 3rd overall - an astonishing achievement. (Author's collection)

The return! 0677 GT, restored and driven by Bob Bodin, in Sienna during the Mille Miglia retrospective of 1984. (Author's collection)

0677 GT as it is today. (Author's collection)

son in 0677 GT which was run by the factory team until Gendebien purchased it for himself in August. He started the season by winning the Tour of Sicily outright on April 14th. His heroic drive in the 1957 Mille Miglia has already been mentioned and the fact that he used 0677 GT to win at Rheims recorded, but it's worth noting that the pair's average speed during the Rheims race was no less than 105.60mph (170kph) for twelve hours. Gendebien and Frere were followed home by no less than four more 250GTs - second, 0607 GT/Phil Hill and Wolfgang Seidel; third, 0683 GT/ Madero and Munaron; fourth, 0509 GT/Papais and Eros Crivellari (who

Incidentally, all these Berlinettas were sold to customers at their ex-factory price (around £6000) in 1957.

Hillclimbs

A word should be said about the European Mountain Championship in which the long wheelbase cars dominated their class of over 2600cc.

The championship included hills in France, Italy, Germany, Greece, Switzerland, Austria, Spain and Por-

was still competing in the Mille Miglia re-run twenty seven years later in another "Tour de France" Berlinetta!); fifth, Camillo Luglio and August Picard in Zagato-bodied 0665 GT.

Luglio took his new Zagato-bodied 250GT to three first places, all in the GT class, plus fifth overall (including second in the GT class) in the Tour of Sicily and sixth overall in the Mille Miglia. Undoubtedly, his best moment of the year must have been winning the Coppa Inter Europa at Monza in front of a home crowd. Not unnaturally, he won the 1957 Italian GT Championship in the over 2600cc class.

On August 4th Wolfgang Seidel led all the way to stave off six 300SL Mercedes to win the GT event immediately preceding the German Grand Prix at the Nürburgring and Curt Lincoln took 0723 GT, the first Berlinetta with covered headights, to win the Swedish Grand Prix, once again defeating the Mercedes 300SLs.

As already mentioned, the Tour de France was won by Gendebien and Bianchi in 0677 GT. However, they were followed home by Maurice Trintignant and August Picard in 0733

GT, another works-entered and run car, and Jean Lucas/Malle, in 0747 GT, to make it a hat trick for Ferrari.

The engine compartment of 0677 GT as it is today. (Author's collection)

Olivier Gendebien adjusts his helmet while talking to his co-driver, Lucien Bianchi, prior to the start of the 6th Tour de France in 1957. They go on to win the event overall.
(Courtesy Kurt Miska collection)

Pontedecimo-Giovi was also won by Camillo Luglio, whilst Aosta-Gran San Bernardo, at the end of the season, experienced complete GT domination by the Berlinettas with Lualdi

tugal and featured such venues as Ollon-Villars in Switzerland, Freiburg in Germany, Mont Ventoux in France, Gaisberg in Austria and, in Italy, Garessio-San Bernardo, Pontedecimo-Giovi and Trento Bondone.

Each hill had to be a minimum of 3.75 miles/6km (though some were up to 22 miles/35.4km long) and with a minimum gradient of 5 per cent. Although there were many well attended hillclimbs in Britain, such as Prescott and Shelsley Walsh, these were too short in length to be included in the European Championship.

In 1957, Mont-Ventoux was the first event of the 1957 championship and the 250GT won immediately with Luglio bringing in the Zagato-bodied 0665 GT first and Lualdi coming second in 0647 GT.

Tour de France, 1957. Trintignant/Picard at the circuit of Rheims in 0733 GT. (Courtesy Van Bever)

winning, Luglio coming second and Eugenio Lubich taking third place in 0597 GT.

Ringoir driving 0707 GT at the Spa-Francorchamps circuit on 25th August 1957. He won the race.
(Courtesy A. Callier)

42

Tour de France 1957. O707 GT at Mont Ventoux hillclimb when driven by Ringoir/Catulle. (Courtesy Maurice Louche Archives)

Towering scenery provides an impressive backdrop to 0707 GT during the 1957 Tour de France. (Courtesy Maurice Louche Archives)

Michel Ringoir driving 0357 GT at Zandvoort in the Grand Prix du MG Car Club. (Courtesy A. Callier)

FERRARI
250 GT
Berlinetta

5

1958

The beautiful lines of 0925 GT, a 1958 Competizione Berlinetta.
(Courtesy H. Mergard)

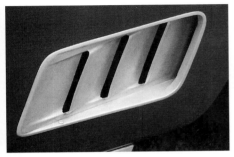

The cooling slots of a 1958 car.
(Courtesy H. Mergard)

The lightweight door handle of a 1958 car. (Courtesy H. Mergard)

The quick-fill "Monza" filler cap of a 1958 car. (Courtesy N. Koel)

The cooling slots of a 1959 car - 1353 GT - whilst under rebuild. Note that this car has an opening for a moveable door (awaiting fitment), which can let cool air into the cockpit. (Author's collection)

For 1958 the bodywork remained very similar to the 1957 second series bodywork, although some cars had a slightly lower roofline.

The three slots behind the side windows became one and most cars had an aluminium strip fitted on the sill below the door line, although some cars had the sills wrapped under the chassis tubes. A three-piece aluminium undershield was fitted to every car.

to assist brake cooling. Gendebien had little trouble in winning again and Bianchi carried out his usual faultless navigation. Another Belgian, Willy Mairesse (of whom much more would be heard), was second in his own 1958 car, 0969 GT. Most photos of Mairesse

Two 1958 Berlinettas, 0911 GT and 1035 GT, meet for a reunion. (Author's collection)

1031 GT has, for many years, been owned by an enthusiast in California. (Author's collection)

Racing in 1958

For the 1958 Tour de France, Gendebien used 1033 GT, a 1958 production car with the body slightly cutaway beneath the sidelights in order

Willy Mairesse's car, 0969 GT, in Momignies, 1958. (Courtesy A. Callier)

through the streets like the Monaco Grand Prix. Taking the first nine places were 250GTs with Gendebien's 0677 GT leading the finishers: Wolfgang Seidel, who had bought 0879 GT, took second place.

At Spa, on May 18th, Seidel was second again in 0879 GT, this time the

driving this car show it with a decidedly battered nearside front wing, so close did Mairesse position the car as he turned into a corner!

Events had really started well in the USA in 1958 when O'Shea/Bruce Kessler won the GT class and came fifth overall in the Sebring twelve hour race with 0773 GT.

The next major event was the three hour race at Pau, situated in France near the Spanish border, an event run

Another continent where it was just as successful as it was in Europe; 0753 GT is driven by R. Cowles in an SCCA meeting at an airfield circuit in Phoenix, Arizona, in 1958.
(Author's collection)

0901 GT driven by Francois Picard during the Nürburgring 1000 kilometre race in 1958. He led the GT class until spinning out at the Karussel.
(Courtesy G. Molter)

winner was Da Silva Ramos in 0749 GT.

Guelfi was third in 0731 GT. Seidel, a sometime member of the factory team when racing a Testa Rossa, had a good year with his own car, placing 1st in races at Trier, Spa and the Eifelrennen.

An Aston Martin DBR1 sports/racing car, driven by Stirling Moss, won the Nürburgring 1000km outright, but a two year old Berlinetta, 0509 GT, driven by Beurlys/Dernier, won the GT class. Picard, driving 0901 GT, a factory-fresh car, crashed whilst in the lead - the car went straight back to the factory for a rebuild. This car had an engine in Testa Rossa tune fitted by

"Beurlys" (Jean Blaton) in 0509 GT winning the hillclimb at Charleroi in August 1958. (Courtesy A. Callier)

the works.

The Mille Miglia was the next event in the racing calendar, held on June 21st-22nd and now transformed into a regularity rally after the previous year's tragic crash of a Ferrari sports/racing car driven by Alfonso de Portago and Edward Nelson. A 1958 Berlinetta, 0911 GT, driven by Luigi Taramazzo and Gerino Gerini, won; second place went to 0597 GT, driven by Villotti and Zampero.

Once again 0677 GT, driven by Gendebien and Frere, won the twelve hours of Rheims on July 6th at record speed, whilst Mairesse and Haldi were second in 0969 GT and Noblet/Peron third in 0619 GT, the Berlinettas filling the first seven places.

Gendebien, as previously mentioned, also won the Tour de France in 1033 GT and the next four places were

Happy smiles from Gomez-Mena, (left) and Meyer (right) on the Promenade des Anglais in Nice before the start of the 1958 Tour de France in brand-new 1035 GT. They did not finish the event.
(Courtesy Maurice Louche)

Victory again! Gendebien, on the right, accepts congratulations, whilst Paul Frere (next but one to his right) explains what the Rheims 12-hours of 1958 was like. Note that 0677 GT has lost its windscreen; the rear screen was also removed to let the slipstream through, as can be seen by the spectators putting their hands inside the car.
(Courtesy Geoff Goddard Picture Library)

Francois Picard waits whilst his mechanics finish work on 0901 GT at Rheims during the 12-hour race on July 6th 1958. He and Bingroff finished fifth overall. Note the blanking tape on the radiator intake and the 'bug screen' on the bonnet. (Courtesy A. Callier)

Rheims again, and this time it's Jean Guichet and Roland Fraissinet in a 1957 car, 0607 GT. They came in sixth overall. (Courtesy A. Callier)

Mairesse at Rheims during the 12-hour race in 0969 GT. (Courtesy Geoff Goddard Picture Library)

1035 GT crewed by Gomez-Mena and Meyer during the 1958 "Tour".

0677 GT waits to be driven for the last time by Olivier Gendebien at the Ollon-Oillars hillclimb in September 1958. Gendebien won his class and then traded 0677 GT in at the factory for 1033 GT the next week.
(Courtesy Graham Gauld)

Quadrio Curzio takes the ex-Lualdi car around Monza during the 1958 Coppa Inter Europa at Monza.
(Courtesy Graham Gauld)

Wolfgang Seidel in 0879 GT at the 1958 Coppa Inter-Europa.   (Courtesy Graham Gauld)

Wolfgang Seidel pilots 0879 GT through the daunting Eau Rouge sweeps on the Spa-Francorchamps circuit in 1958.

Grand Prix du MG Car Club, Zandvoort, July 13th 1958. Michel Ringoir in 0707 GT. (Courtesy A. Callier)

The badge says it all!
(Courtesy N. Koel)

taken by 250GTs.

In the European Mountain Championship, Gendebien took his own Berlinetta, 0677 GT, (bought from the factory in August 1957) to win the Ollon-Villars hillclimb.

Taramazzo drove 0911 GT at Monza to win the Coppa Inter Europa.

The foregoing were the chief events of 1958 and do not include the myriad results of private owners of these cars in club events. Lualdi, for instance, gained five first places in the GT class, whilst Giovanardi entered eight events, always finishing in the top three with 0739 GT. By now the 250GT Competizione Berlinetta in long chassis form was at the zenith of its career, and would see outright victories in competition for only one more year - 1959.

6

1959

Experimentation and development were carried out continually by the factory, particularly with regard to engines. Customers with the best racing records certainly received faster engines. Most engine tuning was confined to making pistons and conrods as light as possible and experimenting with valve timing and compression ratio.

By 1959 the end was in sight for the Competizione Berlinetta in long wheelbase form, because October saw the introduction of the new short wheelbase (2400mm / 94.4 inches) car at the Paris Motor Show. Yet, amazingly, two more body styles on the long chassis appeared in 1959. The reason for the first was simple: Italian lighting law was changed in 1959 when covered headlights were prohibited. Thus, the headlights on the Scaglietti built cars for sale in Italy were moved forward, the perspex covers discarded and chrome surrounds fitted. With this change, the Berlinetta lost its perfection and fluidity of line. Several cars were retro-fitted with covered headlights and most of these received *Tipo* 128C and D engines with the

1461 GT, the NART-entered interim Berlinetta at Le Mans in 1959.
(Courtesy Van Bever)

Left, top: 1335 GT at Monza, 1959. (Courtesy A. Callier)

Left, bottom: 1367 GT, the last Zagato-bodied 250GT built, in 1975 when owned by Jim Boulware of San Francisco. (Courtesy Ed Niles)

1521 GT in the US in the 1970s when owned by Glenn Williams. Note the retro-fitted wing vents. (Courtesy Ed Niles)

sparkplugs mounted on the inside of the cylinder banks.

Pininfarina produced the prototype for the next body style, sometimes called the "Interim Berlinetta." This definition is easy to understand when one compares this body style with the

short wheelbase Berlinetta introduced almost immediately thereafter.

The "Interim" body very much resembled the forthcoming Berlinetta, the difference being that the design was stretched somewhat to cover the

longer chassis. Rounded curves in place of sharp edges seen on previous cars were everywhere, whilst extra small windows appeared behind the side windows. On the prototype and the next car by Scaglietti, a bulge was formed in

The place: Spa-Francorchamps; the date: May 12th 1957. 0427 GT, a 250GT Europa, prepares for the start of a local race.

the bonnet to cover the carburettors, whilst the next five cars, also built by Scaglietti, had open air intakes. 1959 was the first and last time that the long wheelbase cars ran at Le Mans and the new shape cars acquitted themselves well, coming in fourth and sixth overall although a first series upright head-lamp car, chassis number 1321 GT, came third overall to the first and second placed Aston Martin sports/racing cars and won the GT class.

Four of the interim cars had *Tipo*

128D engines but the other three received the new *Tipo* 128D/F engines which now had the sparkplugs mounted on the outside of the cylinder banks. Even so, these engines were as much an "interim" design in engine development as the bodywork that enfolded them. They still retained the three-stud per cylinder head retaining method, even though the conrods were now split horizontally on their centre line and the cylinder heads had twelve ports instead of the old siamesed ports. Coil valve springs now replaced the hairpin variety. Camshafts were still of ten millimetre lift and sixteen studs still retained the alloy camshaft covers. Compression ratio was usually 9.4:1, Borgo pistons were still used, but the distributors were now Marelli ST20 DTEM/A items. These engines were in almost the same state of tune and nearly as powerful as the Testa Rossa unit.

These last seven cars marked the end of the long wheelbase Competizione Berlinetta. Over five exciting years the cars were steadily developed to win races, particularly in the GT class, and succeeded most admirably. Indeed, the long wheelbase cars won more victories than the short wheelbase cars which followed them into production.

Le Mans 1959 and 1461 GT, driven by Andre Pilette and George Arents, dives under the Dunlop Bridge; they finish fourth overall and second in GT. (Courtesy Van Bever)

1321 GT in later life in America, when owned by Manfred Lampe. (Courtesy Ed Niles)

Most of these cars, their competition days over, became road cars: a tribute to their hardiness and durability and nearly all have survived to be swept up in today's collector boom which surrounds the Ferrari 250GT Berlinettas in all forms.

Racing in 1959

The big news for 1959 was that the 250GT Berlinetta made its debut at Le Mans and, whilst Aston Martin DBR1 sports racing cars were first and second, 1321 GT, an upright headlamp Berlinetta driven by "Beurlys" and Elde from Belgium, was third.

Fourth overall was 1461 GT with the "interim" style body; fifth place went to a Spyder California, 1451 GT driven by Bob Grossman and Tavano.

In sixth place was yet another interim Berlinetta, that of Fayon and Munaron (1377 GT). Despite their new bodies, the short noses of the interim cars made them slower down the Mulsanne straight than the older Berlinetta (1321 GT was actually faster than the works Testa Rossa cars on this part of the track).

The 1959 Mille Miglia Rally was won by Carlo Maria Abate and Balzarini in 1333 GT; Abate also came sixth in the GT class of the European Mountain Championship whilst the Coppa Inter Europa was won by Alfonso Thiele in 1389 GT with Carlo Abate second.

Whilst on the subject of Monza, an event called the "Gran Premio Della Lotteria di Monza," which was held in June, should be mentioned. No less than seventeen Berlinettas entered; the sound of all those V12s must have been fabulous!

Thiele also won this event over fifty-two laps, half of them held in the pouring rain. Naturally, twelve Ferraris filled the first places.

The Tour de France ended the season with Gendebien winning again in an interim Berlinetta, 1523 GT.

Carlo Maria Abate in 1333 GT during the 1959 Tour. (Author's collection)

September 20th 1959. 1333 GT at Spa-Francorchamps during the Tour de France, driven by Carlo Maria Abate. He, and his navigator, Balzarini, were third at this circuit and fifth overall on the Tour. The car has an extensive competition history, including first in the Mille Miglia and second in the Coppa Inter-Europa, as well as winning many continental hillclimbs. The car has been in the UK for many years. (Courtesy Ed Niles)

Whoops! A moment during the Lottery Grand Prix at Monza in 1959 when Fritz D'Orey lost control of 0787 GT, but recovered to finish 7th.
(Author's collection)

Lucien Bianchi, now in his own 1033 GT, the car Gendebien won the 1958 event in, had crashed at Monza. Second were Mairesse and Jojo Berger in 0969 GT, and third were Schilde/de la Geneste in 1519 GT, another interim car. Two of these interim Berlinettas had arrived at scrutineering for the "Tour" with disc brakes fitted, but the FIA officials forced the owners to have them removed and the original drum brakes refitted.

In October the new short wheelbase Berlinetta with an all-new body resembling the interim car (but without the small windows aft of the cockpit), an outside sparkplug engine with four head studs per cylinder and disc brakes as standard fittings, appeared at the Paris Salon and the reign of the long wheelbase car was over. Through four seasons of motor racing, both at international and national level, the long wheelbase cars had shown themselves to be a match for, and usually superior to, all opposition.

Jean Blaton at the wheel of 1321 GT at the Cote du Charleroi in May 1959. Blaton was first overall. One month later, Blaton and Dernier drove the car to third overall and first in GT at Le Mans. (Courtesy A. Callier)

Mairesse at the start of the Spa race. (Courtesy A. Callier)

0973 GT seen at Montlhéry in the Paris Grand Prix of 1959, in which it came second in GT, driven by Bourillot.
(Author's collection)

Left: Tour de France 1959 and a squadron of 250GT Berlinettas leave the pits at spa. They are - 158: Schlesser in 1509 GT which did not finish; 171: Dumay/Daboussy in 1521 GT - seventh place; 163: Gendebien/Bianchi in 1523 GT- overall winners; 173: Simon/Thepenier in 0973 GT - did not finish, sump broken by rock; 162: Cotton/Beudin in 1031 GT - did not finish; 164: Lucas/Malle in 1461 GT - crashed. Behind 171 is 161 (1519 GT) driven by Schilde/ De la Geneste which went on to take third place.
(Courtesy Andre van Bever)

Olivier Gendebien's 1959 Tour de France winner, 1523 GT, at Spa-Francorchamps. Gendebien's standing on the pit counter directly behind his car. Note 1523's sandblasted nose!
(Courtesy A. Callier)

The scene at Francorchamps during the Tour de France 1959. Number 170 Abate/Balzarini 1333 GT, and number 163 Gendebien/Bianchi 1523 GT. (Courtesy A. Callier)

*Lucien Bianchi in 1523 GT during
the 1959 Tour de France.
(Courtesy A. Callier)*

*Gendebien driving 1523 GT at Spa
1959; he won the race.
(Courtesy A. Callier)*

250 GT Berlinetta

7

THE DRIVERS

Olivier Gendebien - an interview

Olivier Gendebien, a Belgian who, for many years, has lived in France, was the doyen of Ferrari GT drivers. He was a fast, smooth driver, always particularly good at long distance events.

Paul Frere, the well-respected racing driver and motor racing journalist, once wrote of him - "Gendebien has a particular flair for driving very fast over unknown roads. I sincerely believe that when he finished third overall in a (250) GT Ferrari in the 1957 Mille Miglia, only ten minutes behind Taruffi's winning four litre sports Ferrari, he was the fastest road driver of that time."

Olivier Gendebien won the 24 heures du Mans four times, always in a Ferrari. On top of this achievement, he won the "Tour de France" three times as well as winning the GT class in the Mille Miglia in 1956 and 1957 - all in Ferrari "Tour de France" Berlinettas.

Q: *"Looking at the record, it's obvious that many people believe the Ferrari 250GT was the best GT car ever built. Is this your opinion?"*

A: *"In my opinion, the most successful GT car in 1955 was the Mercedes 300SL but after that came the car I myself was very successful with, the Ferrari 250GT Berlinetta. It's a question of which year you choose."*

Q: *"Just how good was the Mercedes 300SL?"*

A: *"I thought in 1955 that my 300SL was the best there was but, in 1956, I started with the Ferrari 250*

and immediately beat the Mercedes entered in the Mille Miglia."

Q: *"What was your most memorable drive in the long wheelbase Ferrari?"*

A: *"In the Tour de France at the circuit of Pau in 1957. I was about half a lap ahead of Maurice Trintignant, who was in a car with exactly the same specification as mine. The race was run in the rain so engine power did not really count. I stopped just one metre before the finish line to give Maurice the best chance of a result that I could. I could have stayed behind him but I was incapable of driving like that so I went far in front - he couldn't see me and I couldn't see him until he passed me at the very end while I was stationery.*

"1961 was the last time I did the Tour de France. Mairesse was in the pits in the middle of the night, the last night, and we were the last car to start the event. We saw a car in the ditch, we saw the headlamps. Lucien Bianchi, my co-driver, saw it was Mairesse as we got closer and chuckled but to his disappointment I stopped to help Mairesse out of the ditch! This was the only way that Mairesse was able to win the Tour.

"After 1961 I never again drove in the Tour de France because that year Mairesse won by using tricks; he put sixteen inch wheels on for fast circuits so he could go faster, and fifteen inch wheels for hillclimbs, thus changing the overall gear ratio. It was forbidden to change the gear ratios. I protested to the organizing club, the AC de Nice before the start of the rally and they said yes, we know but - and they

shrugged their shoulders - you should not protest because of the spirit of the event, so I said, 'Okay, leave it.'

"At the first event, the Col de Bourse hillclimb, I saw Mairesse's car equipped with the smaller wheels and I then said, 'look, he's is putting on small wheels and then he will put on bigger wheels for the track, like the next race at Spa-Francorchamps.' They [the organizers] said: 'Oh yes, but we'll find a way to stop him.' and so on, but they did nothing. So I said if they allowed people to win a competition which is supposed to be a sport - because I did it as a sport, like every sport I competed in - in this manner I would not enter the Tour de France again."

Q: "So, which do you think was your greatest victory in the long chassis 250GT?"

A: "Every one, everywhere with the 250GT. My most memorable victory with this car was in the Mille Miglia of 1957 where we won the GT class and were third, only eight minutes behind the winning sports/racer which was much more powerful. We won the Grand Prix Nuvolari cup, which I treasure, for this. You know, in my opinion, my four victories at Le Mans were nothing compared with what I achieved at the 1957 Mille Miglia."

Q: "How many times did you frighten yourself?"

A: "At the Mille Miglia? Never. Nor did my navigator, Jacques Wascher, my cousin. You know, it's a question of how much you drive within your own capabilities. I was not like people say in France 'Je ne sais pas motiver' like most of the drivers. I was not motivated because I needed money and glory did not mean much either - that's why I did not take chances in my driving."

Q: "Did your 1956 Ferrari 250GT Berlinetta have many modifications?"

A: "Oh, I don't think so. No, 0677 GT, to my knowledge, was a standard car like all the GTs. But you know, all the Italians, and especially Ferrari, wanted me to win the GT class and beat the Mercedes 300SLs in the 1957 Mille Miglia. That is why Enzo Ferrari put me into the 250GT car of the team instead of the sports racer he had promised me after I won the Tour of Sicily the month before. Then, he asked: 'What car do you want?' I told him I would like the sports car to win the event outright. Three days before the Mille Miglia, Tavoni asked me to come to Maranello and there he said to me: 'We have to win the GT class and I do not think de Portago can do this, so

Willy Mairesse during the 1958 Tour de France at Spa-Francorchamps in 0969 GT.   (Courtesy A. Callier)

you will take the Gran Turismo car.' Then he turned to Portago and said: 'You'll take the sports racer which Gendebien would have had, and I'll be surprised if you go as fast as he will with the Gran Turismo.'

"You know, that is a terrible way to talk to a racing driver. It's the way to kill him and I'm sure that is why Portago died. When we arrived in Brescia I stayed in the Park Ferme until the loudspeakers said Portago had just passed through the last control of Mantova and then I took my own car, a Mercedes 300SL, and I went back to Turin. I was so tired when I arrived there, I went straight to my hotel and the people on the desk were saying 'What a terrible thing' and I said 'What?' and they told me Portago had been

killed ... I was stunned, because he'd such a short distance to go to the finish and I thought he was safe."

Q: "The short wheelbase Berlinetta, was it a lot better than the earlier car?"

A: "I don't know anything about mechanics, all I know is that with the new disc brakes it was a better car."

Q: "Why was the black and yellow stripe on 0677?"

A: "Oh, that's easy. You see, I'm Belgian and sometimes, when I was in the Ferrari in France, they thought I was French. Usually, I live part of the year in France, but I wanted people to know that I was Belgian and so I asked the Scuderia Ferrari at the Tour of Sicily in 1956 - incidentally, the first time I drove the car - to paint the car,

which was naturally painted in red, with two black and one yellow stripe down the middle. Those are the colours of the Belgian flag, you see."

Willy Mairesse

The son of a prosperous timber merchant in a small Belgian town near the French border, Willy Mairesse grew up with no particular interest in motorsport until a friend entered his Porsche 356 in the Liége-Rome-Liége rally in 1953 and took Willy with him as co-driver. Neither had any experience of racing or rallying; indeed, had either done so, they may have chosen a less daunting event on which to cut their motorsporting teeth.

Sure enough, enthusiasm overtook experience and the Porsche came home in a battered state, something that was to happen to Mairesse's cars with sad regularity. The next year, they ventured to enter the Alpine Rally. Same result and Willy's friend retired from motorsport but Willy had found the one thing in life he wanted to do.

He bought a Peugeot 203 and actually completed the 1954 Liége-Rome-Liége Rally, albeit pretty far down the finishing order. For the 1955 event he added a supercharger and finished eighth overall and 1st in class. He was the only competitor, barring the top-rank of well-known drivers, who managed to clear the Vivione pass without any penalty points. He then took the same saloon car and beat all the GT cars to win the twelve hours of Huy, a "Mille Miglia" type event held over secondary roads in Belgium. By now completely bitten by the motor racing bug, Willy sank all his spare cash into a second-hand Mercedes 300SL.

Willy Mairesse entered various hillclimbs and rallies and won every one, excusing his victories by telling people that it was the car which won, not his ability.

At the GT race preceding the German Grand Prix held at the fearsome Nürburgring in 1956, Willy took his 300SL to third place at his first appearance at the track and then really put the cat among the pigeons by beating Stirling Moss, in another Mercedes 300SL, up the Mont Ventoux Hillclimb. One month later, Willy won the Liége-Rome-Liége rally at only his fourth attempt and, truly, the first time he had tried the event in a car capable of winning outright.

To do this, he started his famous rivalry with Gendebien as Gendebien was forced, as a Ferrari factory driver, to accept the heavy Pininfarina-bodied 250GT, 0357 GT, instead of a Scaglietti Berlinetta which he had been expecting. This car could not expect to compete with Mairesse's 300SL.

To finish a very good 1956 season, Willy then won outright the twelve hours of Huy. In the next year's event, however, Willy borrowed another Mercedes 300SL from a dealer and promptly crashed it. He did the same when lent a 2 litre sports racing Ferrari for the Chimay race. However, he did manage to keep his own 300SL intact to win the Spa Grand Prix for GT cars.

Willy hit a bad patch. First he crashed the equipe's Testa Rossa on the first lap of the 24 Heures du Mans and did it again at Silverstone. He then rounded off a bad 1957 by crashing his own 300SL in a tunnel on the Liége-Rome-Liége.

By this time, Willy Mairesse was driving as fast as he, or anyone else, could go and seemed deaf to entreaties to slow down. A case in point was in the Lyons-Charbonniere race of 1958. Willy, taking part in the final stage of a forty mile (64.3km) road circuit, knew that no-one before him had managed to "clean" the stage without penalty. Willie went so fast that he not only "cleaned" the stage, he exceeded the maximum speed the organisers had set!

Jacques Swaters, the owner of Ecurie Francorchamps, helped Willy to buy a 250GT Ferrari Berlinetta (chassis number 0969 GT), and with this he

was second to Gendebien and Paul Frere in the twelve hours of Rheims. He followed this with another second to Gendebien at Clermont Ferrand but then crashed the car on the Tour de France.

In the 1959 "Tour," Willy beat Gendebien at several hillclimbs even though Gendebien was faster at most of the races. Mairesse and Jo-Jo Berger, his co-driver and navigator, were second again.

Enzo Ferrari now stepped into the frame and gave Willy a works drive in the 1960 Targa Florio, where he finished fifth overall and impressed the works team, so much so that he was given several F1 drives: a fact which upset Olivier Gendebien even more as he felt that he had never been given the chance to show his ability in this area. The resentment he felt came to the fore in the 1960 Tour de France where, as already described, he bitterly resented Mairesse's reading of the rules which said nothing about swapping wheels for different events. Willy Mairesse went on to win the Tour with Gendebien in second place and Gendebien, despite having helped pull Mairesse's car out of a ditch, never entered the event again.

For 1961, Willy Mairesse took a works-entered SWB and won at Spa and Clermont Ferrand although finished second to Gendebien and Phil Hill at Le Mans.

On the debit side, Willy came second in the Paris 1000 kilometre race after hitting another car. In the German Grand Prix, Willy went off with only a few laps to go after driving a good

Paul Frere, to the right of Olivier Gendebien, after the duo won the 1958 Rheims 12-hours.   (Courtesy A. Callier)

race. At Pescara, he destroyed his 250GT SWB in a spectacular crash.

For 1962, Ferrari made him test driver and gave him F1 drives: he won the first two races at Brussels and Naples and co-drove the Targa Florio winning car. He then went on to be second at the Nürburgring 1000 kilometres with Mike Parkes, but crashed in the Monte Carlo Rally after posting some excellent practice times. At Spa for the Grand Prix, he didn't get away so lightly and collided with Trevor Taylor's Lotus, finishing up in hospital with severe burns.

Willy Mairesse came back with his tremendous will to win unabated. Over the next few years he was always spectacular; his notable successes included winning the 1966 Targa Florio in a Porsche 906 Carrera.

Alas, poor Willy went off the track in a big way at Le Mans in 1966 while trying to close the door of his Ford GT40. Knowing that his injuries would not allow him to race again, he later committed suicide: a sad end to a tremendous charger.

Paul Frere

Another Belgian; (how they excelled at long distance racing!) After leaving college, Paul Frere became an excellent driver, competing many times for Jacques Swaters' Garage Francorchamps in Brussels. He drove everything from saloon cars to formula one. However, as far as the Ferrari Berlinettas are concerned, Paul shared the winning drive in 0677 GT with Olivier Gendebien in 1957 and 1958 at the Rheims 12 hour race. A man not only interested in the mechanical side of the car's set-up, but also the technique of driving a racing car (he wrote an excellent book on the subject), Paul Frere went on to become a well-known and respected motor racing journalist, most notably forming a very close relationship with Porsche for whom he tested many of their famous racing cars. Today he lives in France.

Phil Hill

The first (and to date only) American to become Grand Prix World Champion (1961). Phil Hill sprang to fame with some notable drives on the west coast of America in an XK120 Jaguar and then a 212 Ferrari Barchetta. Called to

Maranello to become a works Ferrari driver, Phil Hill occasionally drove the 250GT Berlinettas and remembered the "Tour de France" cars as: "Having those awful Houdaille dampers. One lap and they'd be gone. The Koni telescopics on the short wheelbase were a great improvement, but the rest of the [Tour de France] car was typical Ferrari. Engine that ran forever and a chassis you just couldn't break. Great cars, great days." Phil Hill's last international victory was at the Nürburgring in 1967 with the winged Chapparal; shortly thereafter he retired from racing and today runs a thriving restoration shop in Southern California.

Alfonso de Portago

A Spanish nobleman, "Fon" de Portago, who excelled at horseriding and the bobsleigh, gave the "Tour de France" Berlinetta its name, winning the 1956 event with his faithful co-driver, Edmund Nelson. Sadly, only twenty-five miles from the end of the 1957 Mille Miglia, they and eleven spectators were killed when a rear tyre blew out on their Ferrari 335.

8

LIVING WITH A "TDF"

Like most older competition Ferraris, Tour de France Berlinettas have fallen, soared and then crashed (relatively) in value since they were new. It may seem unbelievable now but, in the late 1960s, it was possible to pick up a Berlinetta for £1000 or less. Mind you, you could also have bought a Jaguar C-Type or D-Type for the same sort of money, and £250 could have bought a very presentable XK120. One 14-louvre Berlinetta was actually abandoned on the Hollywood freeway for many weeks!

Today, of course, one would have to spend many times this amount to purchase a good "Tour de France."

March 1965. H. Schumacher, in 0763 GT at the Coupés de Belgique at Zolder, which he went on to win. (Courtesy J. Bellemans)

0767 GT in 1965 when it came second in the Ascoli/Colle San Marco hillclimb. Note the different bodywork. (Courtesy L. Beltrami)

Below: Coupés Benelux at Zandvoort, July 1965. H. Schumacher, in 0763 GT, went for a quick spin! (Courtesy J. Bellemans)

Generally, rising values have been a double-edged weapon, for whilst appreciation encourages owners to go to the expense of having their cars restored to original standard (or better, in most cases), it has also meant that most cars only come out on high days and holidays and very, very few are seen today in competition of any sort.

Above & overleaf: Later life. September 1986 at Zandvoort and 0911 GT awaits the arrival of the author for a Ferrari Owners' Club race. (Courtesy A. Callier)

Thankfully, events such as the retrospective Tour de France and the retrospective Mille Miglia re-runs have ameliorated this situation somewhat.

One sad result of this lack of competitive events has meant that very few people ever get to find out just what these cars are like to drive in anger.

So, what are they like? As an ex-owner, let me tell you. They are *fabulous*! There's no other word for it. They do have their negative points, but we will come to that ...

Right now, let's take a ride in 0911 GT, a car I owned for twelve years. I had her shipped to England in August 1980 and drove her a great deal, doing a mixture of road work and track events. I calculated later that I drove some thirty thousand miles behind 0911's steering wheel. She took on, happily, the feel of a comfortable and favourite pair of old slippers each time I slid into the driving seat. The engine was in bad shape when I obtained the car, but it still lasted eighteen months before it was time for a rebuild. Afterwards, the

difference in the smoothness of running was amazing.

I always made a point of checking the oil and water levels as a matter of course before every trip; it is only sensible as these engines are fast becoming irreplaceable. Then it is open the door and slip into the excellent bucket seat, which has very good thigh support. The only real shortcoming inside the cockpit is the seating position. At six feet tall, I'm too big! In fact, when I wore a crash helmet the top of it was in constant contact with the headlining (I later drove a Drogo-bodied 250GT Berlinetta which was infinitely worse in this respect!) However, leg and arm room are fine, so it's insert the key in the ignition, twist through one hundred and eighty degrees and switch on the fuel pump; said switch being situated beneath the dashboard. When the sound of clacking castanets ceases, push in the key, pumping the accelerator to actuate the accelerator pumps of the three Weber carburettors, and she starts.

It always seems that only two or three cylinders fire at first, the others gradually chiming in as one plays with the throttle. I always tried to hold the engine at a steady 2000rpm for a couple of minutes to let the oil warm up. I say "try" because the accelerator pedal is like a hair trigger and it is very easy for the engine to die. If one removed the bonnet and started up from cold, some fairly spectacular jets of flame would issue from the carburettors occasionally due to the aggressive cam timing.

After two minutes, depress the surprisingly light clutch, engage first gear with a satisfying "clunk" and let out the clutch. Now comes the tricky bit. With a high final drive ratio, (8 x 34), a clutch which is frankly of the "in or out" variety and little torque as one moves off, a juggling act is required. (As an aside, to board Italian ferries with their steep ramps, it was necessary to execute a racing start at 5000rpm, otherwise the car would simply stall halfway up the ramp, much to the amazement of the boarding crew!)

Once underway, everything smooths out.

Accelerate gently away, change up smoothly and watch the oil temperature gauge. It takes about four or five miles (6-8km) for the engine to warm through, and I'm a great believer in letting this happen before starting to use the revs of which the engine is capable.

So, with all gauges registering okay pressures and temperatures, it's press firmly on the accelerator and listen to the note as that wonderful engine reaches 3500rpm - every component seems to be making its own separate noise. Then up to 4500rpm when, with the wild valve timing employed, things really start to happen. Up further to 6000rpm and now there is a continuous howl and one's back is pressed firmly into the seat; it's time to change up once more. Both third and fourth gears select easily into their respective slots in the gate and the car's speed rises relentlessly with wind noise clearly audible through the gaps between doors

An ex-Mairesse Competizione 250GT swb, 2129 GT, as it is today. (Author's collection)

and bodywork.

Now the car is approaching a roundabout and it's time to change down into second gear. The relationship of the brake to accelerator is excellent for heel and toeing and it's a delight to change down through the gears, listening to the "whoop" of twelve cylinders as the throttle is blipped between changes.

The steering is excellent: very light and yet with lots of feel. The handling is also well balanced: just push hard into any corner and the limit is clearly felt when the back tyres start to slide out gently. (Early on in my ownership, I replaced the front anti-roll bar with a stronger one which, paradoxically, reduced understeer). Back off the throttle, apply a touch of opposite lock and the tail swings back into line. The brakes are the weak point: these huge aluminium drums do not fade, but they do require a heavy push to stop the car and they don't like cooling down between applications and nor do they like the wet. (I well remember overtaking a car in the rain and brak-

ing to slot myself behind the car in front when one of the brakes grabbed and I found myself perfectly positioned without having had to turn the steering wheel!) One learns to cope with the brakes' foibles, but it's easy to understand the delight of drivers of short wheelbase Berlinettas when disc brakes were introduced.

That said, it's still a delight to give the Berlinetta its head down dual carriageways or country lanes; it takes everything in its stride.

The ride is firm without being harsh. Doubtless, a few customers who simply wanted a Berlinetta to be in with the "in crowd" may have complained about the hard ride, but they

were people who should have bought a Coupé or Cabriolet, not a GT, the real and uncompromised intent of which was to win long distance road races.

Like all good things any drive must come to an end, and once more the Berlinetta is parked in the garage. Listen one last time to the music of the engine, and then twist the ignition key back through one hundred and eighty degrees. The motor cuts dead as befits its light internal parts. Whenever I walked away from 0911, I always found myself turning back to look at its beautiful shape. The long wheelbase Berlinetta is a superlative Ferrari.

9
RESTORATION

0707 GT as it arrived at Toronto airport from the UK aboard a Boeing 747.
(Courtesy Lance Hill)

Ed Nile's account, in the original version of this book, published in 1984, of the vicissitudes he suffered in having his "Tour de France" (0515 GT) restored proved to be of great interest to readers.

A lot of Ferrari owners wrote to me concerning the original book, and quite a few surprised me by saying how much they enjoyed that particular chapter as they'd gone through something very similar when the time had come for them to restore their own cars. So, for masochists and whoever else *may* be interested, I decided to include another restoration account, this time featuring Lance Hill's car, 0707 GT, a Berlinetta which has had a typically hard life of competition, street use (and abuse!) and occasional track test time.

The restoration of 0707 GT

Raced within two weeks of delivery in the GT race at the Nürburgring, prior to the German Grand Prix of 1957 (where Fangio, in an older Maserati 250F, beat the Ferrari 246 Dinos of Mike Hawthorn and Peter Collins), 0707 GT was the only "TDF" Berlinetta entered that had competed in the event twice already. Unfortunately, it failed to finish in a tough race - but it wasn't for want of trying!

By 1962, 0707 GT, now with a 1958-style covered headlight nose clip, was in England and in a state of complete disassembly. Jock Bruce of Modena Engineering restored the car first, and it was then subjected to a hard life of vintage racing, including the Mille Miglia in 1987 and 1988, before Lance Hill acquired the car (which by then had the correct Alligretti nose clip fitted) in 1992.

Here's what Lance had to say about the nature of the second restoration of 0707 GT - "I was a professional drag-racing driver and engine builder some twenty years ago, have been a senior concours judge with the National Advisory Council for Preservation of the Ferrari Automobile and have long owned Ferraris. I was determined that the old warrior would be returned to the precise state it was in on that summer day in 1957, when Ferrari agent Jacques Swaters presented it, brand-new, to fellow Belgian Michel Ringoir at the Modena Autodromo test track. This meant as thorough and correct a restoration as had *ever* been done on a Ferrari. There would be *no*

The engine being removed prior to restoration. Lance Hill is on the left.   (Courtesy Lance Hill)

compromises, particularly in matters of authenticity and originality."

A great benefit to the restoration was that Jacques Swaters was a friend who had guested at the Hill home in Canada, as was renowned 250GT expert, Jess Pourret. A number of Lance Hill's fellow judges were themselves professional restorers who would unselfishly contribute valuable knowledge and assistance.

Two years were spent scouring the world for documents, old photographs, and original bits and pieces. Even Michel Ringoir was tracked down after months of effort and, astonished and delighted to be asked, recalled the car in remarkable detail.

The work on 0707 GT began with Lance Hill himself doing a comprehensive rebuild of the engine, transmission and rear axle assembly. The rest of the car would be refurbished by a well-known restoration shop.

It was greatly satisfying to find that, without exception, every compo-

continued on page 97

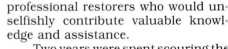

0707 GT on arrival at Lance Hill's restoration shop.
(Courtesy Lance Hill)

An early Berlinetta, 0509 GT, at Lime Rock for an event in 1973.
(Courtesy Ed Niles)

The engine of 0443 GT. Note the
distributors mounted at the front.
(Author's collection)

The second 195 Zagato-bodied Berlinetta, 0537 GT, today.
(Author's collection)

*0677 GT, the ex-Scuderia Ferrari car of Olivier Gendebien. Today, the car is in careful and enthusiastic ownership.
(Courtesy Neill Bruce)*

*0677 GT interior ...
(Courtesy Neill Bruce)*

... engine compartment ...
(Courtesy Neill Bruce)

... and tail, with its distinctive fourteen louvres. (Courtesy Neill Bruce)

The shapely rear of 0767 GT. (Courtesy L. Beltrami)

*Ferrari Owners' Club concours, Castle Ashby in 1995, and a 1959 Berlinetta graces the front lawn.
(Author's collection)*

Piazza Vittoria, 1982, and Eros Crivellari brings 0747 GT to sign on for the Mille Miglia retrospective. Crivellari took part in the "real" event in 1957 in 0509 GT.

0895 GT at Laguna Seca in later years. (Courtesy Ed Niles)

*0805 GT pictured in 1984 sporting a
(very!) non-standard front bumper.
(Courtesy Ed Niles)*

*0907 GT seen in beautiful
condition in September 1984.
(Courtesy Dr M. Kiener)*

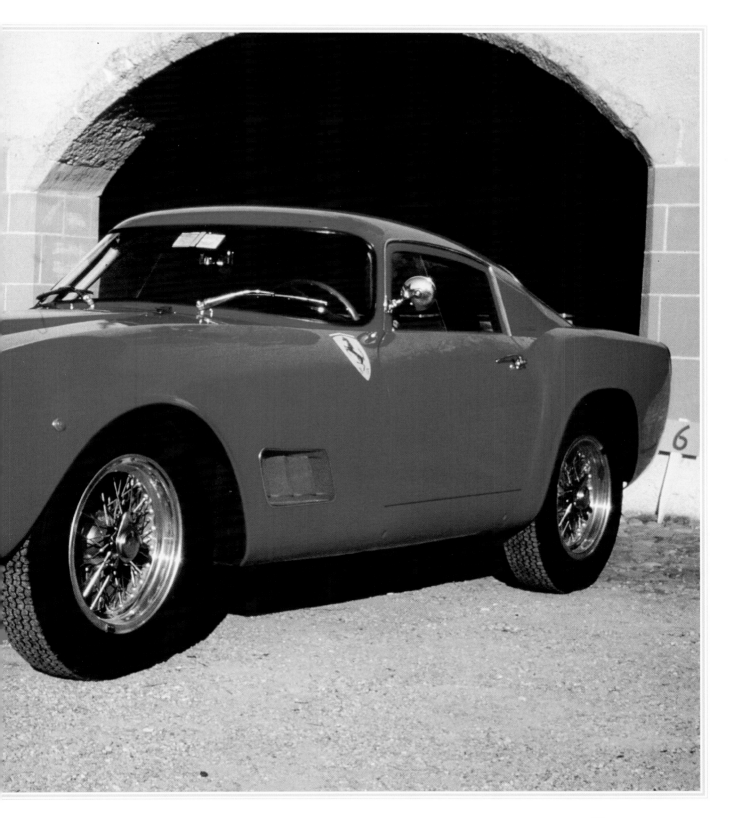

0907 GT

Pages 92/93: 1309 GT, a 1959 car, being raced at Laguna Seca in later years. (Author's collection)

September 1986 and 0911 GT waits to take part in a Ferrari Owners' Club race.
(Courtesy A. Callier)

The chassis of 0629 GT during rebuild. (Courtesy Ed Niles)

1321 GT at Le Mans in 1959. (Courtesy Ed Niles)

1519 GT at La Source hairpin, Spa-Francorchamps in 1959. (Courtesy Dr M. Kiener)

1519 GT at Montlhéry during the 1959 Tour de France when driven by Schilde/De la Geneste. (Courtesy Dr M. Kiener)

The engine of 0707 GT undergoing disassembly. (Courtesy Lance Hill)

nent, from water pump right through to differential case, revealed the same casting and/or stamping numbers as shown on the original factory assembly sheets: an almost unheard of state of affairs with old Ferrari race cars which frequently suffered irreparable damage to such vital pieces.

In the engine rebuild every bearing, seal, bushing and gasket was replaced. New Borgo/AE pistons and rings were fitted. Valve guides were sleeved in bronze and honed to accommodate new stainless steel valves and, in the interests of purity, original-type hairpin valve springs were retained. The antiquated, and notoriously leak-prone block to cylinder head sealing design of steel step rings and separate perimeter gasket, was discarded in favour of one-piece gaskets supplied by Chuck Betz and Fred Peters of California, which are handmade in a modern composite material with soft copper formed around the cylinder and water jacket openings. A modern paper element replaceable cartridge was inserted in place of the old, fixed wire mesh internal oil filter.

Interestingly, in the factory assembly sheet, the original inlet timing of 29/70 and 73/20 had been altered to 45/73 and 42/75 and the exhaust

timing changed from 29/76 and 75/16 to 44/76 and 73/41 before delivery, thus giving an altogether more responsive engine. The *Tipo* 130 10mm camshafts were substituted at the factory for the original *Tipo* 128-types. At the top of the assembly sheet the engine number "65-B" had been crossed out and "87-B" substituted.

The flywheel was refaced, the Fichtel and Sachs pressure plate rebuilt and the clutch plate relined. Following final assembly, the engine was taken outside and fitted to a stand, radiator and gravity-feed fuel lines attached and then the engine was run-in for a total of six hours over two days during which the heads were re-torqued twice and valve clearances reset once. On open exhausts, the sound was glorious!

The original Houdaille dampers had been replaced at some time in the past with Koni telescopics and the grease nipples on the suspension had been deleted. As well as this, the front anti-roll bar was not original. A new,

The paint is chemically stripped from the body.   (Courtesy Lance Hill)

The front suspension of 0707 GT during reassembly.
(Courtesy Lance Hill)

correct anti-roll bar was fabricated as per the original and so, too, were the grease nipples.

All these fittings were trial-mounted prior to stripping the body and chassis in order to ensure a clean and trouble-free final assembly when the time came.

The paint was stripped from the body, which was taken back to bare metal, and the glass windscreen, perspex side and rear windows removed. So, too, was the full interior, including instruments and dash panel, electrical wiring and related components throughout, together with the suspension, brake and fuel systems.

Only then was the skeletal chas-

Front suspension being reassembled.
(Courtesy Lance Hill)

The body after it's been painted.
(Courtesy Lance Hill)

sis and body placed on a rotating jig and every nook and cranny exhaustively examined for damage and corrosion.

Perhaps because 0707 GT had already been apart and carefully preserved for almost half of its existence, there was considerably less work to carry out than is usually found on these cars. Electrolytic corrosion between aluminium body and steel subframes was evident in the customary places such as corners of the bonnet and boot openings, sills, etc; this would be cut out and replaced with new metal but, in all, corrosion was minimal.

The chassis was degreased, stripped, bead blasted, carefully inspected, primed and repainted. Brake master and slave cylinders were hot tanked, and sleeved or honed as required. Brake backing plates and drums were stripped, beaded and refurbished, whilst new flexible brake hoses were acquired and steel brake lines fabricated.

The original aluminium undershield panels were missing so new ones were fabricated using photos and the undershields of 0585 GT as patterns. Ringoir remembered the car having a cold air box for the carburettors, so another was made.

The wiring harness was laid out and duplicated precisely as lengths, colours, connections, terminals, sleeves and bindings. The steering wheel was reconditioned, retaining the original wood. The instruments were

rebuilt, as were the switches, relays, fuse blocks and panel with the dashboard being painted crackle black as per the original.

The steering box was rebuilt and the radiator recored, whilst the header tank was boiled to remove scale and the filler neck replaced. Exhaust manifolds were refabricated and a new exhaust system fitted.

The outer rims of the wheels had been polished at some time in the past, obliterating the "Carlo Borrani" name and number "RW3264," although the date of manufacture "11/56" was clear on the inside rim. The wheels were stripped, outer rims trued on a lathe,

restamped and then repainted.

All the suspension components were stripped and magnafluxed. Two front wishbones had hairline cracks and were weld-repaired and heat-treated to original specifications. Components were straightened as required, rebushed and prepared for plating and, afterwards, furnace-baked to release hydrogen embrittlement.

The rear leaf springs were refinished, rebushed, new nylon friction barriers installed and original job numbers repainted. The front springs were replated.

The inner wing panels were replaced as required with original style solid body rivets used. Fibreglass mat was bonded to the inside of the front

Rear brake and suspension during reassembly. (Courtesy Lance Hill)

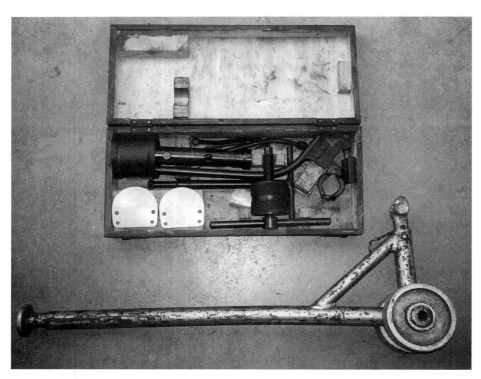

wings for rigidity and to provide protection from stones thrown up by the tyres. Five original Englebert 600 x 16 tyres were located from various obscure, dusty, far-flung corners of the world and mounted to be used for show purposes only.

A set of Houdailles, not set up for 0707 GT, required some 10 hours each of machining and modifications, after which they were run-in and adjusted on a purpose-built rig.

The fuel tank was too corroded to be saved and so it was replicated, the original filler neck and vent tube being re-used.

Great care was taken to ensure that the correct colour and type (Wilton wool) carpet was fitted, together with the original diamond-type vinyl used around the base of the cockpit. The seats and door panels were re-upholstered as per the original cream vinyl.

Various parts, such as the rubber driveshaft coupling, handbrake cable, throttle linkage mountings and bearings, ignition coils, voltage regulator, horn, compressor and relay, plus lights and the interior mirror, were replaced or remanufactured.

There's a photograph of evening maintainance being done on 0707 GT during the 1958 Tour de France in which one of the factory racing jacks can be seen in the boot. These were rare then and now nearly impossible to find, so reproductions were made. However, in keeping with the strict dedication to authenticity, an original was located and acquired with the assistance of Jess Pourret.

Even more rare are the original engine overhaul tools that came in a small wooden box that the factory mechanics used at race sites. Reproductions of these have been made as well but, again, only a complete and original set would do and, eventually, one was found.

The car was assembled in its entirety with everything tested, aligned and adjusted as required. 0707 GT

The engine bay of 0707 GT awaiting refitting of the motor.
(Courtesy Lance Hill)

0707 GT on the lawn at the 1996 Pebble Beach Concours, restoration complete. (Courtesy Lance Hill)

was completed in time to be shown at the 1996 Pebble Beach Concours d'Elegance at Monterey, California.

The restoration of 0911 GT

I obtained my own Tour De France (0911 GT) in 1980 and my ownership experiences are probably typical.

At first my sheer delight at owning the car blinded me to its faults, but a rising cloud of oil smoke from the engine, which made its way into the cockpit, soon made me realise that attention to the engine could not long be avoided. A broken head gasket forced me into action and so, being fortunate enough to live near Forward Engineering, a very good engineering shop, I left the car with them, with the instruction to investigate possible problems and let me know.

The 'phone call from Ron Beaty, head of Forward Engineering, came that afternoon. "Do you want the good news or the bad news?" he asked. "I'll have the bad news first" I replied. "The

heads have been machined like door wedges and all the pistons have been touching all the valves" said Ron. "Oh ... what's the good news?" Ron paused, then said "You don't need new sparkplugs!"

So the decision was taken to remove the engine and investigate further. This was in February 1982 and I had already entered the car in the Mille Miglia re-run set for May of that year, so speed was called for. Upon stripping the engine further, it became apparent that years of neglect had taken their toll: the heads had been progressively milled down until the only choice was to get new heads (very difficult!) or make copper gaskets to fit - the latter course was chosen. The required parts included pistons, rings, valves, guides and bearings. Here, Graypaul Motors came to the rescue. I visited them with my list of wants and, to my amazement, they could supply everything. Having acquired the parts I took them to Dave Butcher, Ron's most experi-

enced engine builder who was charged with completing the engine in time for the Mille Miglia.

I arrived at Forward Engineering on the day of completion of the engine to find an army of blue-overalled mechanics swarming on, around and under the car. The engine rebuild was finished. With exhortations to stay below 4000rpm for the first 500 miles, my wife and I set off for Italy. There were no problems whatsoever and we had the time of our lives, arriving home with the car in good order. In particular, the engine sounded beautiful.

But then came trouble. Two months after arriving home I discovered to my horror that the crankshaft was cracked. Greypaul once again came to the rescue. For a (comparatively!) reasonable cost they arranged to have a crankshaft made for me by Farndon Engineering. The finished article was magnificent. Once more Dave Butcher was prevailed upon to screw it all back together and now the oil pressure is

60-80lbs/sq inch when hot. It does give one confidence!

The next step was the gearbox. It was obvious the synchromesh cones were very worn, so off the gearbox went to a local specialist in Redditch who split the box using made up special tools. A visit to Jeff Hales of Greypaul Motors resulted in new synchromesh cones (again, in stock) so the gearbox was back in the car inside two weeks - outstanding service, I thought.

The car's aluminium bodywork was next on the list. An examination of it showed that, apart from the nose (which had been rebuilt to look like one on a Monza after the car went off the road in the 1959 Mille Miglia), the bodywork was in remarkably good condition and did not need much work.

The same could not be said of the brakes and suspension. The problem with the brakes was that when they were used hard in conditions such as circuit testing, or in a race, they really worked very well, but in normal everyday use on the road they were, quite simply, inadequate. Upon stripping the brakes little was found to be at fault, and so a softer brake lining - supplied by Ferodo - was substituted for the extremely hard ones fitted. This made a big difference when road driving but also brought to light the suspension defects: under braking the car would swerve to left and right, as well as crashing over bad road surfaces. I knew the suspension was supposed to be firm, but this did not seem right at all ...

So, off came all the suspension, which is easier than it sounds as the rear springs, suspended on a shackle at each end, proved to be a) built like the Forth bridge and b) almost immovable after 26 years on the car! Eventually, with lots of WD40 and elbow grease, they did come off. If you perform this task yourself, beware! The energy stored in these nine leaf springs curls them into U-shapes, but luckily they just missed the mechanic's head! The front suspension came apart fairly simply, but removing the phosphor-bronze bushes from the chassis entailed pressing a steel rod behind each bush and a well-swung hammer blow. When the coil springs were removed we were amused to find that one side was one coil shorter than the other! Probably, a spring had broken at some time in the past and the assembly had been put back together without a replacement spring.

A local engineering firm made new bushes and pins, whilst new coil springs were also made. I took this opportunity to fit a stronger front anti-roll bar and the difference in 'feel' and handling is amazing. To complete matters I fitted Dunlop racing tyres of the period, sizes 600 x 16 inch front and 6.50 x 16 inch back. I know that Engelberts or Pirelli Corsa were the 'official' tyres of the period, but as these are now all but unobtainable I opted for the Dunlops.

Finally, the interior. I was fortunate that 0911 GT still had its quota of Veglia instruments, together with original diamond quilt padding beneath the instrument panel and around the pedals which, apart from cleaning, was left alone. The headlining and remainder of side and door trim was also original, so this was also well scrubbed (after removing it from the car) and replaced.

The carpet and seats were very tatty so new carpet - using the remains of the old - was made up in black and the seats completely remade and trimmed in tan leather to match the rest of the interior. Some rewiring was done at the same time.

Finally, in the autumn of 1984, all was completed. It was with a very satisfactory feeling that I drove the car out of the garage for a 20-mile test run. All now seemed perfect. The ride and handling - not to mention the brakes - complemented each other and I could savour the experience of driving on the road a truly representative 250GT "Tour De France."

250 GT Berlinetta

10

THE "TOUR DE FRANCE" BERLINETTA IN MODERN COMPETITION

Like most historic (or just plain "old," depending on how you see things) competition cars, the "Tour de France" Berlinettas were hard-used and abused in their first life but, with the advent of well mannered cars, were pressed into service as fast road cars when their competition days were over. This was a role for which they were well suited and, in 1960, when the short wheel-

1127 GT in later life, still competing. (Author's collection)

historic racing, have, nevertheless, reappeared on the scene.

Most of them, being essentially base car had superseded the model on the tracks, there can have been few cars on the roads of Europe and

1321 GT at a Ferrari Club of America meeting in 1975. (Author's collection)

This page: 0911 GT at Zandvoort in 1986. (Courtesy A. Callier)

0911 GT, with the author driving, at Donington in September 1989. (Author's collection)

America that could match their sheer pace. Even the then new E-type Jaguar would have had a hard struggle against a 250GT "Tour de France" Berlinetta on the road!

Naturally, as the values of the old Ferraris diminished during the 'sixties and 'seventies, a lot of "bodging" went on when it came time to make repairs. But, when historic racing started in the 1970s, loyal owners put their cars into proper shape and went pot-hunting again. Over the years several cars have been well "tweeked" with the substitution of disc brakes for original drums, telescopic dampers for the old Houdaille lever-arm type and exten-

0911 GT at the Isle of Man meeting in 1989. (Author's collection)

sive engine modifications being common.

Not unnaturally, in classic sporting events, the long wheelbase car found itself once again at a disadvantage to the SWB and GTO. However, in the early 1980s, the Mille Miglia was revived as a retrospective event for cars built up to 1957 and, suddenly, all the Berlinettas came out to play again!

Even better, the Tour de France was later also revived as a retrospective event, giving two prestigious outings per year to owners who wished to enjoy their cars in the manner for which they had been designed.

250 GT
Berlinetta

Introduction and disclaimer
First of all, many thanks to Jess Pourret who wrote the definitive history of all the 250GT Berlinettas in his wonderful book *Ferrari 250GT Competition Cars*. I have used his individual chassis histories as a basis for the following register, which includes all the information I have collected over many years with help from other enthusiasts such as Mike Sheehan and Lance Hill. Particularly where the older cars are concerned, there are still gaps which I have been unable to fill. **Important: by the very nature of these cars' histories, their age, potential value and sheer number of sources of information, it is more than likely that there are omissions, mistakes and/or inaccuracies in the individual** histories of particular cars. **Do not regard the information recorded here as proof of authenticity or provenance of any particular car, nor as a record of its complete history.**

Although a "delivery date" is given, this is actually the date of the factory invoice to the first customer. The actual delivery date could vary widely from the invoice date.

Dates of events affecting each car, such as when it was raced or sold, are noted where known. As many photographs as possible of the individual cars have been included, but note that these are of variable quality.

Naturally, most of these cars were used in competition and, naturally again, they were crashed (sometimes many times), rebuilt,

Below and pages 107 & 108: 0707 GT assembly sheet.

Telaio tipo	508 B20	matricola	0707 GT
Motore tipo	128 B20	matricola	0707 GT
Cambio tipo	508 B20	matricola	12 B20
Ponte tipo	508 B20	matricola	298 D
Collaudo il		Consegnato il	
Cliente			

modified and developed. A lot of the cars have been rebuilt and, in some cases, re-created. The reader, if tempted to buy one of these magnificent cars, should **always** seek impartial expert advice and adopt the motto *Caveat Emptor!* (Buyer Beware!)

Autotelaio tipo 508 B20   Motore matricola N° 0707 GT   Telaio matricola N. 0707 GT

FOGLIO MONTAGGIO AUTOTELAIO

Trasmissione 508B/350 con giunto ant.Fabbri ø 80 e post.SAGA

Freni   ant.508B/452 post.508B/436 tamburi AZ

Mozzi   ant.166/70856/857   post.342/70865/866

RUOTE 16x5,50 RW 3264 cromate

Ammortizzatori ant.508/64457 post.508/61106

Sterzo Scatola guida 508/325 R=1/18 Volante 508/329 Guida sinistra

Radiatore acqua e olio   508B/81072

Serbatoio carburante tipo 508B/810046 modificato   capacità lt.137

Indicatore di livello   SI

Marmitta di scarico   ABARTH doppie

Batteria   Baroclem

NOTE

MOTORE N° 65 B

CAMBIO N° 12 B20 comando centrale tipo 342

PONTE N° 298 D R=8x32

Frizione tipo FICHTEL & SACHS

Balestre posteriori 508B/61337

Molle anteriori 508B/64553 Kg.395 fless.0,237 %.

Freni ant.con suole 553 incollate,cilindretti ø 28 anelli paraacqua.

Freni post.   "   554   "   "   ø 28

Carrozzeria Scaglietti con volante Nardi

Silentblok per biscottino baleste da 65 Shore.

Serbatoio carburante modificato capacità lt.137 anzichè lt.111

Strumenti BORLETTI in Km: Ta 265; Te 078; Mn 286; IL 205; Cg 008/c.

Finita il

Data 22/6/57   Il Capo Reparto   Franchini

Autotelaio tipo 508 B20   Motore tipo 128 B20   Matricola N. 0707 GT

FOGLIO MONTAGGIO MOTORE N° 65 B 87 B.

Basamento 128B20/10591 Aq Coppa olio 128/11962 Aq.9 Filtri

Albero motore 128B/12251 bancata ø 60

Pistone BORGO rif.3492   Rapp. 9,2   Peso   gr.228

Anello tenuta 1 torsionale + 1 conico Raschiaolio 2 con bisello magg.ø 73x4x

Bielle 125/14102   Peso gr.405   Pompa acqua 125/16300

Teste cilindri 128/17070 con canne TRIONE   Coperchi 128/17019

Guarnizioni teste Klingerite 1000,anelli rame amianto,cerchietti in acciaio.

Valvola asp. tipo 400/16135   Valvola scar. tipo 225/16676 a/c 130

Molla richiamo valvola ø 400/16148   Alberi distribuzione 128/17053/054 alzata

Scatola distribuzione   128/22404

Pompa mand. olio 128B/24552   Pompa di recupero   NO

Pompa benzina 2 FISPA   Filtro benzina con le pompe   Pres aria alta FAT 410:

Carburatore tipo WEBER 36 DCL/3 N 3

Accensione con **spinterogeni** Tipo ST 100 DTEM/E Fase: D

Frizione tipo F & S   Carico

Giuochi Albero-motore 0,05   Punteria 0,15   0,2

		AA 20/20	AA 20/80
		CS 28/20	CS 28/18
		42/75	73/41
	45/73		44/76

PRIMO MONTAGGIO

Data inizio montaggio   Data fine montaggio   Montatori   Guerzoni / Cavazzuti

OSSERVAZIONI

Raschiaolia con bisello maggiorato.
Bronzine di banco TRIONE e di biella Vandervell.

Data 25/5/57   Il Capo Reparto   Franchini

OSSERVAZIONI DEL PRIMO RODAGGIO

Motore regolare stà bene sotto carico,pressione nel carter 35/40 % H2O
a pieno regime.

MOTORE DELIBERATO 29/5/57

Il Capo Reparto   Taddei

SECONDO MONTAGGIO

Data inizio lavoro   Data fine lavoro   Montatori

OSSERVAZIONI

Data   Il Capo Reparto

```
Autotelaio tipo    508 B20      Motore matricola N  0707 GT      Telaio matricola N  0707 GT

                    FOGLIO MONTAGGIO AUTOTELAIO

Trasmissione   508B/350 con giunto ant.Fabbri ø 80 e post.SAGA

Freni         ant.508B/452  post.508B/436 tamburi AZ

Mozzi XXXXXX  ant.166/70856/857  post.342/70865/866

XXXXXXX  RUOTE  16x5,50  RW 3264 cromate

Ammortizzatori  ant.508/64457  post.508/61106

Sterzo  Scatola guida 508/325 R=1/18 Volante 508/329 Guida sinistra

Radiatore acqua e olio          508B/61072

Serbatoio carburante tipo    508B/810046 modificato        capacità lt.137

Indicatore di livello           SI

Marmitta di scarico      ABARTH doppie

Batteria         Baroclem

NOTE

MOTORE N° 65 B

CAMBIO N° 12 B20 comando centrale tipo 342

PONTE N° 298 D R=8x32

Frizione tipo FICHTEL & SACHS

Balestre posteriori 508B/61337

Molle anteriori 508/64553 Kg.395 fless.0,237 %.

Freni ant.con suole 553 incollate,cilindretti ø 28 anelli paraacqua.

Freni post.   "    554   "      "     ø 28

Carrozzeria Scaglietti con volante Nardi

Silentblok per biscottino baleste da 65 Shore.

Serbatoio carburante modificato capacità lt.137 anziché lt.111

Strumenti BORLETTI in Km: Ta 265;Te 078;Mn 286;IL 205;Cg 008/c.

Finita il

Data  22/6/57                     Il Capo Reparto    Franchini
```

0357 GT

THE FIRST 250GT COUPÉ. Delivered to Cosheen in Switzerland. 1954 car, but not invoiced until March 5 1956. Retained by the factory for first 30,000 kilometres. Pininfarina alloy-bodied coupé. Job number: 13441. *Tipo* 508/112. Registration: 1 R 612. 1270kg. Introduced at the Paris Motor Show of 1956. Black with red top.

1956:	5/3/ Sold to Ecurie Francorchamps.
1956:	17-23/9/: Tour de France. 3rd OA, 3rd in GT class. Race number: 74. Gendebien/Ringoir.
1957:	Lyon-Charbonniere Rally. DNF. Gendebien. Race number: 124
	Sold to Ecurie Francorchamps, Belgium.
	Sold to Ringoir in Belgium.
1960:	Sold to G. Focquet in England.
1978:	Sold to the Hilton brothers.
	Restored by Vic Norman.
1979:	Sold to Teddy Pilette in Belgium.
1992:	For sale at SS Imports of Alabama.
1996:	For sale whilst undergoing restoration.

0369 GT

Invoiced 3/5/55 to Recordati in Italy. Pininfarina competition Berlinetta. Job number: PF13939. *Tipo* 508/112. Grey with red interior. Registration number: RE 27836.

1958:	Sold to C. Chaplin.
1958:	24/11; Sold to A.C. Ferrari. Registration number: MO 49525.
1958:	Factory records say: "destroyed". 19/12/58. Automobili Club Italia records confirm this.

0383 GT

Invoiced 5/7/55 to Coulibeuf in France. Pininfarina competition Berlinetta. Job number: 13446. *Tipo* 508/112. 1050kg. 600 x 16 wheels. Light grey/blue. Registration: GE 91309.

1955:	Mille Miglia. DNS. Coulibeuf/Behra.
1956:	Sold to F. Bastiglio di Remo, Italy, for Paulo Lena to drive.
1956:	Coppa Inter Europa, Monza. 3rd OA. Race number: 84. Paolo Lena.
	4/11: Coppa Serravalle-San Marino. 2nd OA. P. Lena.
	17-23/9: Tour de France. DNF. Race number: 176. Lena/Palanga.
1956:	Sold to Chinetti, USA.
	Sold to Greenspun.
1957:	23/3: Sebring twelve hours. Race number: 16. DNF. Gendebien/Greenspun.
1958:	Sold to Frank Adams.
	Watkins Glen. DNF. Crash. Adams.
1959:	Lime Rock. DNF. Crash. Adams.
1963:	Sold to McLellow minus engine. Burned and buried after garage fire.
	Exhumed, remains sold to Bill Pierce.
	Sold to T. Myr.
1986:	For sale US $50,000. Restored in Florida.
1988:	Sold to Glenn Kalil.
	Sold to John Collins (Talacrest) in England.
1995:	August: Monterey Laguna Seca races. Race number: 107. Richard Mattei.
1996/7	Now in Sweden.

Note: A replica of this car has been built upon a 250GTE chassis.

0385 GT

Invoiced 4/4/55 to L. Bertlett in Italy. Pininfarina comp. Berlinetta similar to 0369 GT but with opening boot. Silver. *Tipo* 508/112. Pininfarina job number: 13447. Displayed at the 1955 Turin Motor Show.

1956:	Sold.
1960:	Sold to Crepaldi in Milan.
1962:	Sold to G. Cappe in Milan.
1964:	Sold to L. Seraidaris of Greece.
1965:	Bad crash.
1975:	Sold to Steve Barney in USA. To Modena for restoration.
1980:	Restoration finished, painted silver. 5 50%
1986:	Sold to Brian Brunkhorst.
1986:	Sold to HartmutIbing.
	Sold to Kato in Japan.
1993-95:	Offered for sale by John Collins of Talacrest (red).
1995:	Sold to Harry Levantos, UK.

0393 GT

Invoiced 14/7/55 to Dubonnet in France. Pininfarina comp. Berlinetta with louvres on bonnet and finned rear wings and flat covered headlights. First 250GT Berlinetta with the new coil spring front suspension. *Tipo* 508/112. P/F job number: 14974. "1955 Berlinetta Tipo MM azzura." 214bhp at 6500rpm.

1955:	Paris Auto Salon for display in October by Paul Vallee of Autoval.
1956:	24 Heures du Mans, DNS. Dubonnet/Trintignant.
1963:	Sold to Richard De Jagger through Garage Moloff.
1966:	Sold to Robert Spencer.
1980:	Car totally disassembled in USA and offered for sale at $15,000.
1980:	Sold to Dr Ronald Mulacek, La Grange, Illinois. Restored.
1990:	Shown in Illinois concours.
1995:	Sold to Dennis Machul.

0403 GT

Invoiced 15/10/55 to J.G. Murray in Italy. 2480mm wheelbase. Pininfarina Comp. Berlinetta with rear bodywork similar to the 375MM of Ingrid Bergman and with flat covered headlights. P/F. Job number: 14975. Registration number: MO 37284.

	Sold to C. Bross.
	Sold to J. L. Steele.
	Sold to Anthony Bamford.
1978:	Sold to Marvin Johnson in USA. Restored.
1985:	For sale at $175,000.
1989:	Sold at auction for $1,400,000.
1990:	With Daniel Frankhauser, Kusnacht, Switzerland.
1993:	With Peter Fandel, Germany.
	Sold to Roschmann, Augsburg.

0415 GT

Invoiced 16/11/55 to Perdisa. 2480mm wheelbase. Pininfarina Comp. Berlinetta. P/F. Job number: 14976. Louvres on bonnet. The last LWB with a 250MM type body.

1955:	Nassau GT race, 1st OA. Race number: 23. Derujinsky.
1956:	This could be the car with which Gendebien won the Giro di Sicilia on 8/4/56. Note that the inset headlights of this car are completely different to the protruding, chrome-surrounded lights of 0503 GT.
	17/6: Oporto GP, fastest lap and 1st OA. De Portago.
	Sold back to factory, Ferrari Works test driver Sighinolfi killed in accident in the car.
1960:	Sold to Don Beach in USA, re-engined with Chevrolet V8.
1974:	Sold to Mike Curley.
1983:	For sale with Rudi Pas of Belgium, £80,000, completely rebuilt with Ferrari V12 inside plug engine number 0445 GT.
1983:	Sold to HartmutIbing, Germany.

1984:	Sold to H. Gericke.
1993:	Sold to Dieter Roschmann.
1993:	T de F Retro. Roschmann/Roschmann.
1993:	Mille Miglia Retro. Roschmann/Roschmann.
1994:	Mille Miglia Retro. Roschmann/Roschmann.
1995:	Mille Miglia Retro. Roschmann/Roschmann.

0425 GT

1955 car. Invoiced 20/4/56 to Wax and Vitale in Italy. Competition Berlinetta. PininfFarina job number: 12490. 2480mm wheelbase. 1956 Geneva show car. Fourteen louvres on rear panels. In 1985, Ed Niles and the author found the car still in the ownership of Chester Bolin (Ed had sold him the car in 1974). Thankfully, Ed managed to buy the car which was in original - if dilapidated - condition and have it restored to pristine condition. Registration number: UEV 431.

1958:	Sold to J. Haynes in USA.
1974:	Sold to Chester Bolin in Orange County, California.
1986:	Sold through Ed Niles, completely restored.
1990:	Sold to Tanaka in Japan.

0443 GT

Invoiced 12/2/56 to Jean Estager in France. Pininfarina coupé in alloy. *Tipo* 508/112. 1160kg. 600 x 16 wheels. Registration number: BO 65270.

1956:	6-12/7: Rally des Alpes. 5th OA, 1st GT. Estager/Pebrel.
	Perthuis Hill. 1st in class. Estager.
	17-23/9: Tour de France. DNF, Race number: 71. Estager/L. Rosier.
1957:	24-28/4: Acropolis Rally. 1st OA, 1st GT. Estager/Estager. Race number: 7
	Lyon-Charbonniere Rally. DNF. Race number: 112. Estager.
	Sold to Ferrari, sold to Weprin, California, USA.
1972:	R. Jordan, USA.
197?:	Sold to Chuck Weber, California.
1985:	Sold to Laurence Auriana, NY.

0503 GT

Invoiced 8/5/56 to Dr Augusto Caraceni of Italy. Scaglietti-bodied Competition Berlinetta with extra windscreen wiper above screen, sliding windows, warm air induction and no bonnet blister. *Tipo* 508B-513/128. Rebodied in 1958 with Scaglietti Berlinetta body with open headlights. Today, restored to original configuration. Registration number: MO 53672.

1956:	8/4: Tour of Sicily. 4th OA and 1st in GT. Race number: 254. Gendebien/Wascher. (Note: this may have been 0415 GT or 0509 GT. Note headlights).
	28-29/4: Mille Miglia. 5th OA, 1st in GT. Race number: 505. Gendebien/Wascher.
1957:	Sold to E. Simeone "Kammamuri."
	12/5. Mille Miglia. 15th OA. Race number: 442. E. Simeone "Kammamuri"/Bellini.
	1/9: Garessio Colle San Bernardo. 2nd OA. E. Simeone "Kammamuri".
	15/9: Coppa delle Cimino. DNF, E. Simeone killed. Sold back to Ferrari and rebodied.
1959:	Sold to Z. Tchkotoua.
	Faucille Hill. 2nd in GT class, Tchkotoua.
	Tour de France: DNS. Race number: 167. Tchkotoua/Testut.
	Pontedecimo-Giovi. 3rd in GT class, Tchkotoua.
1962:	Sold to G. Ledford, USA.
	Sold to T. Thomson.
1980:	Sold to Joe Marchetti.

1984:	Sold to Michael Leventhal.
1984:	Mille Miglia retrospective. No: 185. Leventhal/Rabin.
1985:	Sold to Robert Rubin, NY.
1986:	Mille Miglia retrospective. Rubin/Pirrone.
1987:	Sold to European Auto Sales.
1988:	Sold to Reiner Simon of Switzerland.
1989:	For sale at $1.5m. Sold to Robert Gregory, Dallas, Texas.
1990:	Mille Miglia retrospective. Gregory/Reilly. Car was crashed in this event and rebodied with a 1956 Berlinetta, original style, by Bob Smith Coachworks.
1992:	Mille Miglia retrospective. Race number: 295. Gregory/Foley.
1992:	Pebble Beach Concours. 1st in class M2.
1992-95:	Stored in Switzerland.
1995:	Tour de France Auto retrospective, Race number: 115. Gregory/Reilly. Engine no: 0401 GT?

0507 GT

Delivered on 23/4/56 to Olivier Randaccio in Italy. Scaglietti Comp. Berlinetta. Red. No bonnet blister, flat covered headlights, sliding windows. *Tipo* 508/128. Registration number: MO 304826.

1956:	28-29/4: Mille Miglia. DNF. Randaccio.
	1/9: Garessio Col San Bernardo, 3rd OA. Race number: 326. O. Randaccio.
1957:	12/5: Mille Miglia. DNF. Race number: 326. O. Randaccio.
	8/9 Coppa Inter Europa. 8th OA. Randaccio.
	Coppa Lombardi. 2nd OA. Randaccio.
	29/9: 20th Pontedecimo-Giovi. 3rd OA. Randaccio.
1958:	15/6: Varese-Campo de Fiori. 3rd OA. Randaccio.
	Aspern Hill. 4th OA. Randaccio.
	7/9: Monza, Coppa Inter Europa. 1st in GT? Race number: 75. O. Randaccio. Sold to Rene Trautman, France.
1959:	To Scaglietti's where modified with a lower nose, covered headlights and a smaller grille.
	Urcy Hill. 8th OA. Trautmann.
	St. Antonin Hill, 1st. Trautmann.
	Monza, Coppa Inter Europa. DNF. Trautmann.
	Gineste Hill. 1st in GT, Trautmann.
	Trophees de Provence. 1st in GT. Race number: 126. Trautmann.
	Macon Hill. Race number: 87, result unknown. Trautmann.
	Tour de France. DNF, Race number: 160. Trautmann/Gele.
1960:	Sold to Switzerland and bodywork modified front and rear.
1968:	Sold to C. Ahlfeld of Denmark.
1994:	In Aalborg Automobilmuseum, Denmark.

0509 GT

Delivered on 27/4/56 to G. Giovanardi of Italy. Scaglietti Comp. Berlinetta. Red. *Tipo* 508B/128B. 241bhp at 7000rpm. 1050kg. Flat covered headlights, wind-up windows, spare wiper motor above windscreen. Registration: 39311 MO, BO 99399, 928 RO.

1956:	28-29/4: Mille Miglia. 24th OA, 8th in GT. Race number: 448. Giovanardi/Meier.
	24/6: Vermiccio Rocca di Pappa, 1st in GT. Giovanardi.
	15/7: Bologna - San Luca. 2nd OA, 1st in GT. Giovanardi.
	22/7: Aosta - Gran San Bernado, 3rd OA. Giovanardi.
	16/9: Sorrento San Agata, 2nd OA, 1st in GT. Giovanardi.
	7/10: Tre Ponti-Castelnuovo. Result unknown. Giovanardi.
	28/10: Modena Autodromo, Campianato Emilia Romania, 1st OA. Giovanardi.
	Bolzano-Mendola. 3rd in GT. Giovonardi.
1957:	31/3: Corsa dell Toricelle Hill, 3rd in GT. Giovanardi.
	Sold to Olivier Papais of Venice.
	24/2-1/3: Sestrieres Rally. 8th OA, 2nd in GT. Papais/Crivellari.
	12/5: Mille Miglia. 16th OA, 6th in GT. Papais/Crivellari.
	13/14/7: Rheims 12 hours, 4th OA. Race number: 86. Papais/Crivellari.
1958:	March. Sold to Jacques Swaters, Ecurie Francorchamps.
	4/5: Dinant Hill. 1st in GT. Blaton.

11/5: Herbeumont Hill. 1st in GT. Blaton.
1/6: Nürburgring 1000kms. 1st in GT, Race number: 41. Blaton/Dernier.
5/10: Namur GP, 1st in GT. Blaton.
Spa 1000km. DNF. Crash with frontal damage. Blaton.

1958:	Sold to Bob Dusek in USA.
1992:	Mille Miglia retrospective.
1992:	Still with Bob Dusek.

0513 GT

Delivered to V. Collocci in Italy on 27/3/56. Scaglietti Comp. Berlinetta. No blister on bonnet. Wind-up windows. 1030kg. *Tipo 508/128*. Red. Registration number: BO 69214.

1956:	8/4: Giro di Sicilia. 3rd in GT. Colloci.
	17-23/9: Tour de France. 3rd OA. Race number: 72. Trintignant/Picard.
1957:	Sold to Ferrari.
	Sold to Chinetti.
	Sold to Greenspun. Registrastion number: N-34.
	22/6:Elkhart Lake 500 miles. 1st in Sport class. Race number: 16. P. Hill/Greenspun.
	5/7: Lime Rock. 1st OA. Greenspun.
	8/12: Nassau. Race number: 68. P. Hill/Greenspun.
1958:	22/3: Sebring.
1960:	Sold to W. Bowman, Chevrolet V8 engine fitted.
1974:	Sold to M. Curley.
1981:	Sold to B. Taylor. Fitted with the engine from 0515 GT.
1982:	Sold.
1989:	Engine from 250GT Ellena 0803 GT installed.
1990:	For sale.

0515 GT

Delivered to Vladimiro Galluzio on 30/6/56 in Italy. Zagato "Double bubble" bodywork. Cost: 900,000 lire. Rolling chassis from Ferrari for 3,000,000 lire. *Tipo 508B/128B*. Bonnet blister, extra wiper motor above screen. Registration number: MI 312234. Blue with white roof. Registration number: GE 100136. Sold on 2/7/56 to SILA, Galluzzi's Company.

1956:	6/7: Coppa Dolomiti. 2nd in class. Galluzi.
	Coppa Sant Ambroeus. Galluzzi.
	Coppa d'Oro. 5th OA. 2nd in GT. Galluzzi
	29/7: Giro di Calabria. Race number: 853. Galluzzi.
	Coppa Inter Europa at Monza. DNF. Race number: 94. Galluzzi.
1957:	22/2: Sold to Scuderia St. Ambroeus.
	July: Sold to Orlanda Palanga, Genoa
	Sold to Luigi Taramazzo of Imperia, Bordighera.
	1/9: Garessio-San Bernardo. 1st OA. Taramazzo.
	8/9: Coppa Inter Europa. 3rd OA. Race number: 80. Taramazzo.
	14/10:Pontedecimo-Giovi. 2nd OA. Race number: 328. Taramazzo.
1958:	14/9: Molino-Cocconato Hillclimb. 1st OA. Paulo Lena.
1959:	Sold to J. Mazzo, Sold to Malago, sold to Goldoni.
1960:	Sold to Ed Niles in USA and bought and sold six times (McQuid, Shiffer, Strader, Van Dyke, Brinker, Scott) by Niles.
1973:	Sold to J. Sullivan.
1970's:	Sold back to Niles, sold to B. Pessin.
1983:	Sold back to Ed Niles, restored with original motor by Steve Tillack, Won "Best Ferrari" prize at Pebble Beach.
1985:	Registration number: 374 JEV.
	Sold.
1993:	Sold to Lorenzo Zambrano of Monterey.
1994:	Entered in Monterey historic races and Pebble Beach Concours by Pablo Gonzales.

0537 GT

Delivered to SILE for Cornelia Vassali, (Camillo Luglio's wife) on 4/6/56 in Italy. Zagato "Double bubble" Berlinetta similar to 0515 GT except stripped interior, seat bolted directly to floor. Dark grey with red on bonnet blister. Registration number: GE 91344.

1956:	20/5:Trofeo Sarda. 3rd OA, 1st in GT. Luglio (result credited to Luglio, even though before invoice date).
	7/7: Bolzano-Mendola. 3rd OA, 1st in GT. Luglio.
	8/7: Coppa d'Oro delle Dolomiti. 5th OA. Luglio.
	29/7:Giro di Calabria. 3rd OA, 1st in GT. Luglio.
	7/9: Monza. Coppa Inter Europa. 2nd OA. Race number: 90. Luglio.
	21/10:Roma GP. 2nd OA. Race number: 18. Luglio.
1957:	14/4: Giro de Sicilia. 5th OA, 2nd in GT. Luglio.
	1/5: 5th Trofeo Bruno and Fofi Vigorelli and Coppa Constantini. 1st in GT. Luglio.
	12/5: 24th Mille Miglia. 6th OA, 2nd in GT. Race number: 441. Luglio/Carli.
	30/6: Mont Ventoux Hillclimb. 1st in GT. Luglio.
	8/9: Monza, Coppa Inter Europa. 1st OA. Race number: 83. Luglio.
1960:	Sold to P. Nobili.
	Sold to USA.
1962:	Sold to Ed Niles.
	Sold to D. Janpol.
1970:	Sold to Druker.
	Sold to C. Nassiry.
1992:	Mille Miglia retrospective. Albrecht Guggisberg.
1994:	Sold to Robson Walton in USA.
1996:	Restored by Skip McCabe for Pebble Beach Concours. Now has leather interior.

0539 GT

Delivered on 28/6/56 to Edouardo Lualdi Gabardi in Italy. Scaglietti Comp. Berlinetta. Red. *Tipo* 508/128. Registration: VA 50036. Flat covered headlights, wind-up windows.

1956:	7/7: Bolzano-Mendola. 2nd in GT. Lualdi.
	22/7: Aosta-Gran San Bernardo. 2nd in GT. Lualdi.
	19/8: Trapani-Monte Erica. 1st in GT, Lualdi.
	26/8: Selva di Fasano. 1st in GT. Lualdi.
	23/9: Trento Bondone. 1st in GT. Lualdi.
	7/10: Treponti-Castelnuovo. 2nd OA, 1st in GT. Lualdi.
	14/10:Pontedecimo-Giovi. 2nd in GT. Lualdi.
	4/11: Serravalle-San Marino. 1st in GT. Lualdi.
	Coppa San Marino. 1st in GT. Lualdi.
	Trofeo Montagna (Trophy of the Mountains), 1st in GT over 2600cc. Lualdi.
1957:	Sold. (Note: This *could* be the car sold to "Lorica" which won the Rally Jeanne d'Arc on 16th June. Race number: 163.
1990:	Sold to Fabrizio Violati.
1990:	Mille Miglia retrospective. DNS. Confidati/Confidati.
1991:	Mille Miglia retrospective. Sandias/Nociti.
1992:	Mille Miglia retrospective. Marano/Piacquadio
1993:	Mille Miglia retrospective. Asakawa/Hada.
1994:	Mille Miglia retrospective. Sakamoto/Takashi. Race number: 254.

0555 GT

Delivered to Ferraro on 8/9/56 in Italy. Scaglietti Comp. Berlinetta. *Tipo* 508/128. Registration number: CT 50279. Wind-up windows, light con-rods (436 grams), flat covered headlights.

1957:	12/5: Mille Miglia. 7th OA, 6th in GT. Race number: 431. Ferraro alias "Hippocrate."
	6/10: Trieste Opicina Hill. 5th OA, 2nd in GT. Ferraro.
	Targa Florio. DNF. Clutch. Ferraro.
1958:	Sold back to Ferrari in May.
	Sold to S. Le Pira.
	Messina Colle San Rizzo. FTD. Le Pira.
	15/5: Monte Pellegrino Hill. 3rd OA, 1st in GT. Le Pira.
	Coppa Nissena. 2nd OA. Le Pira.
	7/9: Monza. Lottery GP. 8th OA. Le Pira.
1959:	Targa Florio. 9th OA, 1st in GT. Race number: 120. Le Pira/Todaro.
	Avola-Avola Hill. 1st OA. Race number: 154. Le Pira.
	Coppa Nissena. 2nd OA. Le Pira.

Targa del Busento. 1st OA. Le Pira.
Catania - Etna. 3rd OA, 1st in GT. Le Pira.
Monte Pellegrino. 3rd GT. Le Pira.
1960:   Coppa Belmonte. 1st. Le Pira.
Coppa Nissena. 3rd OA, 1st in GT. Le Pira.
Targa del Busento. 1st OA. Le Pira.
Sold to G. Gasso.
1961:   Pellegrino Hill. 1st GT. G. Gasso.
Catania-Etna. 1st in GT. Race number: 448. G. Gasso.
Valdesi Santa Rosalia. 2nd OA, 1st in GT. G. Gasso.
1962:   Sold to USA. Dick Merritt/G. Wales/Andrea. Sold to Kennedy.
1972:   Sold to Previtt. Sold to W. Dickey.
1976:   Sold to Thoroughbred MotorCars.
1985:   Sold to Jo Marchetti. Sold again.
1986:   Sold to Leventhal.
1986:   Mille Miglia retrospective. Leventhal/Leventhal.
1987:   Mille Miglia retrospective. Leventhal/Leventhal.
1995:   Colorado Grand. Leventhal/Leventhal.

0557 GT

Delivered to Alfonso de Portago on 10/9/56. 1020kg. Grey. *Tipo* 508/128. Wind-up windows and radio! Registration number: BO 69211, SPY 250, GR7307.

1956:   17-23/9: Tour de France. 1st OA. Race number: 73. De Portago/Nelson.
21/10: Rome GP. 1st in GT. De Portago
1957:   7/4: Montlhéry, Coupés USA. 1st OA and GT. Race number: 14. De Portago.
(Note: This could be the car used by "Lorica" to win the Rally Jeanne d'Arc on 16th June, Race number: 163).
Sold to Keith Schellenberg of Eigg in Scotland.
1983:   Sold to Paul Palumbo.
1993:   Sold at Brooks Auction to Zambrano, Monterey.
1994:   Restored by Bob Smith Coachworks.

0563 GT

Delivered on 10/9/56 to Giacomo Peron in France. 1100kg. French Racing Blue. *Tipo* 508B/128B. Wind-up windows, 600 plus 650 x 16 wheels. Registration number: 214813 TO.

1956:   17-23/9: Tour de France. 8th OA. Race number: 75. Peron/Bertramier.
Montlhéry, Coupés du Salon. 2nd OA. Race number: 25. Peron.
1957:   Rallye des Forets. 1st in GT. Peron.
7/4: Montlhéry, Coupés USA. 1st in GT. Race number:8. Peron.
Rallye Printemps. 2nd OA, 2nd in GT. Peron.
13-14/7: Rheims 12 hours. DNF. Race number: 84. Peron/Lucas.
Montlhéry. Paris GP. 1st in GT. Peron.
30/6: Rallye de L'Allier. 1st OA. Peron.
4/8: Corsa de Razal. 1st in GT. Peron.
Montlhéry, Coupé du Salon. Race number: 17. Peron.
Armagnac Rally. 3rd OA. Peron.
16-22/9: Tour de France. 5th OA. Race number: 171. Peron/Bergraf.
1958:   Pau 3 hours. 7th OA. Race number: 61. Peron.
27/4: Planfoy Hill. 1st in GT. Peron.
16/3: Rallye des Forets. 1st in GT. Peron/Peron.
Lyon-Charbonnieres Rally. DNF. Race number: 128. Peron.
Routes du Nord Rally. Race number: 6. Peron.
Sold back to Ferrari, sold to B. Kessler in th USA.
1960:   Sold to R. Wakeman.
1973:   Sold to Larry Taylor. Crashed and rebuilt.
1983:   Sold to R.Gent. Restored by Joe Piscassi and Tom Sellby with matching handmade luggage.
1996:   Still with Richard Gent.

End of the 1956 model year Berlinettas

0585 GT

Delivered to Tony Paravano in USA on 15/11/56. First 14-louvre Berlinetta. Registration number: 11496. Red with blue and white stripe. Dark blue interior and factory fitted oil filter.

1957:	4/4/57: Palm Springs. Race number: 103. Skip Hudson. Disqualified as not a "production model."
195?:	Sold to Marty Arrounge, USA.
1965:	Sold to Richard van der Water, USA.
1955:	Sold to the Disney Studios for use in *The Love Bug*.
1960:	Sold to Mrs Harvey Schaub.
	Sold to Mike Schaub. Abandoned on the Hollywood freeway for several weeks.
1977:	Sold to J. Rothman in the USA.
1979:	Sold to Bruce Lavacheck, Phoenix, USA. 75,039 kilometres indicated.

0597 GT

Delivered to Eugenio Lubich in Italy on 1/1/57. 14-louvre Berlinetta. Red. Registration number: VE 34837, PA 55162, EMO 340C.

1957:	12/5: Mille Miglia. 43rd OA, 11th in GT. Lubich/Villotti.
	1/9: Aosta -San Bernardo, 3rd in GT. Lubich.
1958:	21-22/6: Mille Miglia Rally. Race number: 3. 2nd OA, Villotti/Zampero.
	8/6: Coppa Citta Asiago. 1st. Villotti.
	2-3/8:Giro Di Calabria. 2nd OA, 1st in GT. Zampero.
	6/7: Bolzano/Mendola Hill. 7th OA, 3rd in GT.
	7/9: Coppa Inter Europa. Race number: 79. Zampero.
	Sold to Todaro.
1959:	Coppa della Sila. 1st OA. Race number: 174. Todaro.
	Rome, Trofeo Betsoja, 1st OA. Todaro.
	Coppa Sila de Fasano. 1st OA and in GT. Todaro/Trapani.
	Monte Erica. 2nd OA, 1st in GT.Todaro.
	Coppa Nisena. 1st OA, 1st in GT. Todaro.
	Avola/Avola. 2nd GT. Todaro.
	Monte San Pellegrino. 4th OA, 2nd in GT. Todaro/Valaesi.
	San Marino. 2nd OA, 1st in GT. Todaro.
	Passo di Ragano/Belle Campo. 9th OA, 1st in GT. Todaro.
1960:	Buenos Aires 1000km. 9th OA, 1st in GT. Todaro/Munaron.
	Venezuela GP. Todaro/Munaron.
1963:	Sold to Mark Rigg, UK.
1976:	Sold to K. Knight.
1983:	Restored and engine rebuilt by Terry Hoyle.
	Sold to Chris Mann.
1993:	Mille Miglia retrospective. Mann/Mann.
1995:	Sold to Richard Thwaites.
1997:	Nose now being corrected to proper shade.

0607 GT

Delivered to Cavazzoli of Italy on 2/2/57. Red. Scaglietti Comp. 14-louvred Berlinetta. Extra wiper motor above windscreen, wind-up windows. Registration number: MI 335318, 8116 BG 13.

1957:	14/4: Giro di Sicilia. 4th OA. Munaron.
	12/5: Mille Miglia. DNF. Race number: 426. "Madero."
	Sold to Wolfgang Seidel of Germany.
	26/5: Nürburgring 1000km. Crashed in practice by Von Trips.
	13-14/7:Rheims 12 hours. 2nd OA. Race number: 88. Seidel/P. Hill.
	4/8: Nürburgring. German GP support race. 1st OA. Seidel.
	4/8: Renania/Palatinat GP. 1st in GT. Seidel
	16-22/9: Tour de France. DNF. Race number: 172. Seidel/P.Hill.
	Sold to Jean Guichet.
1958:	7/4: Pau 3 hours. 4th OA. Race number: 58. Guichet.
	13/4: Lavande Rally. 1st in GT. Guichet/Souchon.
	4/5: De la Gineste Hill. 1st in GT. Guichet.

11/5: St. Antonine Hill. 1st in GT. Guichet.
Auverne 3 hours. 7th OA, 6th in GT. Guichet.
6/7: Rheims 12 hours. 6th OA. Race number: 76. Guichet/Fraissinet.
Rally du Petrole. 1st. Guichet/Bobin.

1959: Mille Miglia Rally. 2nd OA. Guichet/Happel. Race number: 7.
       Sold.
1967: Sold to Thuysbaert.
1969: Sold to Pozzoli in France.
1974: Sold to B. Comte.
1987: Mille Miglia retrospective. Comte/Comte.
1988: Mille Miglia retrospective. Comte/Comte.
1992: Tour de France retrospective. Comte/Comte.
1992: Tour de France retrospective. LaBaume/Moreau.
1994: Tour de France retrospective. LaBaume/Moreau.
       Sold to Claude Fernandez.

0619/0805 GT

Delivered to Autoval in France on 28/3/57 for Pierre Noblet. A 1956 Scaglietti Comp. Berlinetta with wrap-around rear window, bonnet blister, new type distributors, 1/20 steering ratio and no louvres on the sail panel. Medium grey. Re-numbered 0805 GT in December 1957. Registration numbers: 8665 FZ 75, AH 3806, NF 6536.

1957: 16/6: Rally Jeanne d'Arc. 1st OA. "Pertin" Noblet.
       16-22/9: Tour de France. 8th OA. Race number: 163. "Pertin" Noblet/"Lorica" Cavrois.
       Traded in at Modena. Renumbered (0805 GT) in December.

1958: Sold to K. Kaufmann, Caracas,
       Sold to Chimeri.
       11/5: Coppa Citta di Valencia. 1st in GT. Chimeri.
       14-15/6: Vuelta Aragua-Maracay. 1st OA. Chimeri.,
1959: To Modena to be rebodied with 1959 upright headlight body and an SWB-like bonnet plus bumperettes. Shipped back to Venezuela.
       Sold to M. Mierklik.
1961: Sold to Florida, USA.
1970: Sold to C. Previtt.
1971: Sold to A. Brand.
1972: M. Derish.
       Sold to D. Scoby.
1974: Wide World of cars.
       Sold back to D. Scoby.
1975: Sold to A. Pedretti.
1976: Sold to Fazzano.
1981: Sold to Tom Davis.
1982: Sold to Cavallino Collectors cars.
1990: Mille Miglia retrospective. Thiebat/D'Alpaos.
1993: For sale by Maranello Rosso Collection, Italy.

0629 GT

Delivered to Paola Lena of Italy on 6/2/57. A 1957 14-louvre Scaglietti Comp. Berlinetta. *Tipo* 508B/128B. The air door in the front wing is set lower than in the other cars with this type of bodywork. Registration number: GE 96010, MI 340452. VC.44247. The author was in Los Angeles in 1983 when Bradley Balles finished the restoration of this car and took a ride with Bradley. The car looked and felt superb.

1957: 24/2-1/3: Sestrieres Rally. 6th OA, 1st in GT. Race number: 20. Lena/Palanga
       14/4: Giro de Sicilia. 6th OA. P. Lena.
       16-22/9: Tour de France. DNF. Crash. Race number: 164. Fabregas/Soler.
1958: Sold to Carlo and Massimo Leto di Priolo, Milan.

15/5: Vienna Aspern race. 1st in GT. Race number: 30. C. Leto.
7/9: Monza, Coppa Inter Europa. 2nd OA. C. Leto.
14/9:Stallavena-Boscochiesanuova Hill. 1st OA. C. Leto.
5/10:Innsbruck Airfield. 1st in GT. M. Leto.
Kranenbittern Airport. 1st in GT. Race number: 73. C. Leto.

1959: GP Lotteria di Monza. DNF. Leto.
Sold to Riggenberg of Switzerland.
Mitholz-Kandersted. 3rd in GT. Riggenberg.
St Ursanne les Rangiers. 1st in GT. Riggenberg.
Schauinsland Hill. 1st in GT. Race number: 96. K. Stangl.

1960: Sold to USA. T. Lockwood, D. Bacon, Car has no running gear.
1976: Sold to Bradley Balles.
1983: Car completely restored with Ferrari parts.
1996: Still with Bradley Balles.

0647 GT

Delivered to Edouardo Lualdi Gabardi of Italy on 7/3/57. Scagletti Comp. Berlinetta with fourteen louvres on the sail panel. Red. Registration: 53490 VA, 26140 BZ.

1957:
1/9: Aosta-San Bernardo Hill. 1st in GT. Lualdi.
31/3: Corsa delle Torricella. 1st in GT. Lualdi.
28/4: Bologna-San Luca Hill. Ist in GT. Lualdi.
5/5: Coppa della Consuma. 1st in GT. Lualdi.
2/6: 3rd Coppa Lombardia. 1st. Lualdi.
18/8: Trapani-Monte Reice Hillclimb. 1st. Lualdi.
1/9: Aosta-San Bernardo Hill. 1st in GT. Lualdi.
8/9: Coppa Inter Europa, Monza. 2nd OA. Lualdi.
15/9: 9th Coppa del Cimino. 1st. Lualdi.
Mont Ventoux hill. 2nd in GT. Lualdi.
6/10: Trieste-Opicina Hill. 1st in GT. Lualdi.
13/10: 4th Coppa L. Carli, Monza. 1st in GT. Lualdi.
10/11:Campianato Sociale Scuderia Arena. 1st GT. Lualdi.
Trofeo della Montana. 1st in GT over 2600cc. Lualdi.

1958: Sold back to Ferrari.
Sold to Quadrio Curzo.
27/4:Coppa San Marino. 1st in GT. Race number: 324. Curzio.
4/5: Bologna san Luca. 2nd OA, 1st in GT. Curzio.
7/9: Coppa Inter Europa, Monza. DNF. Race number: 78. Curzio.
Sold to P. Helm in the USA.

1967: Bad crash with Helm on Hollywood freeway. Car sold to Ed Niles, broken up and parts sold. Chassis left at S & A Italia Sports Cars.
1989: Chassis and some parts discovered and sold to a buyer in England. Car being rebuilt.

0665 GT

Delivered to Cornelia Vassali in Italy for Camillo Luglio to drive on 10/6/57. The third Zagato-bodied "Double bubble" Berlinetta. Flat covered headlights, extra wiper motor above screen. Black/silver. Registration numbers: GE 127879, MI 379254, GE 97979.

1957:
13-14/7: Rheims 12 hours. 5th OA. Luglio/Picard.
1/9: Aosta San Bernardo. 6th OA, 2nd in GT. Luglio.
8/9: Coppa Inter-Europa, Monza. 1st OA. Race number: 83. Luglio.
29/9: 20th Pontedecimo-Giovi. 3rd OA, 1st in GT. Luglio.
Italian GT Champion over 2600cc.

1958: Pau 3 hours. DNS. Luglio.
Sold to Vladimir Galuzzi.
21-22/6: Mille Miglia Rally. DNF. Race number: 8. V. Galluzi.
Rimini. San Marino. 5th OA. V. Galuzzi.
6/7: Campiano/Vetta d'Enza. 2nd OA. V. Galuzzi.
Coppa Sant Amroeus. Race number: 151. V. Galluzi.
Sold to Crepaldi.

Sold to V. Corradini.
1959:   Sold to J.C. Meade, USA. One race.
1967:   Sold to E. Marshall.
1972:   Sold to W. Wright. Restored.
1976:   Sold to Rick Milburn.
1978:   Sold to Joe Marchetti.
1979:   Sold to Peter Kaus in Germany.
1996:   Still with Peter Kaus.

0677 GT

Scuderia Ferrari team car. Sold to O. Gendebien on 28/8/57. Scaglietti fourteen-louvre Comp. Berlinetta. Engine 128MMC of 3117cc capacity used for Testa Rossa experimentation but with three, instead of six, carburettors. *Tipo* 508B/20/128B20. Engine tested 5/4/57. 245bhp at 7000rpm. Extra wiper motor above screen. Maroon with yellow stripe. Registration numbers: MO 49. BO 81831, BO98155, MI 429195.

1957:   14/4: Giro de Sicilia. 1st OA. Race number: 315. Gendebien/Wascher.
        12/5: 24th Mille Miglia. 3rd OA, 1st in GT, 1st in Nuvolari Cup. Race number: 417. Gendebien/Wascher.
        13-14/7: Rheims 12 hours. 1st OA. Race number: 80. Gendebien/Frere.
        Sold to Gendebien.
        16-22/9: Tour de France. 1st OA. Race number: 170. Gendebien/Bianchi.
        6/10: Montlhéry, Coupés du Salon. 1st in GT. Race number: 19. Gendebien.
1958:   7/4: Pau 3 hours. 1st OA. Race number: 56. Gendebien/Bourillot.
        6/7: Rheims 12 hours. 1st OA. Race number: 164. Gendebien/Frere.
        27/7: Auvergne 3 hours. 4th OA. 3rd in GT. Race number:
        76. Gendebien.
        31/8: Ollon-Villars. 1st in GT. Gendebien.
1959:   Sold back to the factory. Sold to G. Gerini.
        Tour de France. DNS. Race number: 169. Gerini.Meier.
        Lottery GP, Monza. Race number: 55. Gerini.
        Targa Florio. 12th OA. Gerini.
        Sold.
1963:   Sold to T. Fowler in USA.
        Sold to Kelly.
        Sold to Newman.
        Sold by Ed Niles.
1973:   Sold to Pessin.
1976:   Sold to A. Woodall.
1977:   Sold to R. Bodin. He had the car sympathetically and lovingly
        restored to its 1957 Mille Miglia condition.
1984:   Mille Miglia retrospective. Bodin/Bodin.
1986:   Mille Miglia retrospective. Bodin/Bodin.
1987:   Mille Miglia retrospective. Bodin/Dedolph.
1989:   Mille Miglia retrospective. Bodin/Bodin.
1990:   Colorado Grand. Bodin/Bodin.
1994:   Sold to P. Vestey. Car dismantled and restored again by D.K. Engineering and R.S. Panels.
1995:   Mille Miglia retrospective.

0683 GT

Delivered to O. Capelli in Italy on 19/4/57. Scaglietti fourteen-louvre Comp. Berlinetta. Extra wiper motor above screen. Red. Registration number: MI 350541.

1957:   12/5: 24th Mille Miglia. 25th OA, 8th in GT. Race number: 439. Cappelli.
        16-22/9: 6h Tour de France. DNS. Race number: 173. Madero/Munaron.
        13-14/7: Rheims 12 hours. 3rd OA. Race number: 90. Madero/Munaron.
        Pau 3 hours. 3rd OA. Race number: 52. Madero/Munaron.
        Castelquarto-Venasca. Madero.
1958:   Coppa Weiss-Marchal. Modena. 1st OA. Race number: 20. Madero.
1959:   Sold to A. Zampero.
1960:   Sold to Venezuela.
1989:   Fabrizio Violati.

0689 GT

Delivered to De Micheli in Italy on 29/3/57. The fourth Zagato-bodied Berlinetta. No "Double bubble"roof, covered headlights, tandem master cylinder and extra wiper motor above screen. *Tipo* 508B/128B. Registration number: F1 98172, 116892 FI, TA 19747.

1958:   29/6: Predappio Rocca della Caminate. 5th OA, 2nd in GT. Di Micheli.
         6/7: Campiano Vetta d'Enza. 1st in GT. Di Micheli.
1959:   Coppa Rimini-San Marino.
1960:   Sold to U. Filotico.
         Agano-Cappella. 1st in GT. U. Filotico.
         Trofeo Venturi. Frascati-Toscolo. 1st in GT. U. Filotico.
         Coppa Fagioli. 1st in GT. U. Filotico.
         Coppa d'Oro, Modena. Race number: 154. U. Filotico.
         Selva di Fasano. 2nd OA. U. Filotico.
1962:   Sold to Nub Turner, Michigan in USA.
1974:   Sold to K. Hutchinson
1979:   Sold to John Hadjuk, Sold to Gary Rice of Lancing, Michigan.
1981:   Sold by Joe Marchetti.
1982:   Sold to Charles Glapinski.
1986:   Mille Miglia retrospective. Glapinski.
1987:   Mille Miglia retrospective. Glapinski/Glapinski.
1989:   September. Sold to Van der Meene, Holland.

0703 GT

Delivered to Albino Buticchi of Italy on 7/5/57. Scaglietti fourteen-louvre Comp. Berlinetta. *Tipo* 508B20/128B-57B. Later a 1958-style covered headlight nose was fitted.

1957:   12/5: 24th Mille Miglia. 9th OA, 4th in GT. Race number: 430. Buticchi.
1960's: Sold to Curzio
         Sold to Delameter in the USA and then Norman Silver. At some time fitted with 375MM type nose and tail lights.
1977:   Sold to FAF in Atlanta.
1990:   Colorado Grand, John Apen. GTE engine installed.
1996:   Still with John Apen.

0707 GT

Delivered to Garage Francorchamps in Brussels, Belgium for Michel Ringoir. Invoiced July 19th, the last Scaglietti Comp. Berlinetta with fourteen louvres made. Red/Cream interior. *Tipo* 508B/20. Originally fitted with *Tipo* 128B-65B engine giving 220bhp at 7000rpm but later updated to a *Tipo* 87B engine giving 245bhp. Returned to Scaglietti's 8/57 for fitment of limited slip differential and seat modifications. Crashed in 1960, converted to a 1957/58 covered headlight nose and damaged right hand door. Engine replaced with a Jaguar 3.4 litre unit. Door remade in steel. (FOC Magazine letter, Volume 8, Paul Schouwenburg). Now rebuilt to proper shape nose. Larger than normal fuel tank (137 litres), no fuel gauge, only low-level indicator light. First car to be equipped with *Tipo* 130 10mm lift camshafts. Lightest conrods of any TdF, 405 grams. Road tested by Paul Frere - 0 - 100 kph in 7.6 seconds; 0 - 160 kph in 15.2 seconds; standing kilometre in 26.9 seconds; standing quarter mile in 15.0 seconds.
1957:   4/8: Nürburgring. GT race, race number: 41. 6th OA. Ringoir.
         25/8: Spa-Francorchamps RACB GP 1st in GT. Race number: 35. Ringoir.
         1/9: Brussels 1500 metres sprint. 1st in GT. Ringoir.
         16-22/9: Tour de France. DNF. (Camshaft drive). Race number: 166. Ringoir/Catulle.
         5/10: Namur Hillclimb. Race number: 7, 1st GT. Ringoir.
         6/10: Namur Hillclimb. Race number: 78, 1st GT. Ringoir.
         20/10: Dinant Hillclimb. Race number: 0, 2nd OA. Ringoir.

1958:   9/2: Routes de Nord Rally. Race number: 7. 1st OA, Ringoir/Blondiau.
        23/3: Cote D'Houyet. Race number: 69, 1st OA. Ringoir.
        24/3: Cote D'Houyet. Race number: 67, 3rd GT, Ringoir.
        7/4: Pau 3 hours. 8th OA. Race number: 54. Ringoir.
        27/4: Knokke, Flying Kilometre. Race number: 110. 3rd OA, 1st in GT.
         Ringoir.
        27/4: Knokke. Flying Kilometre. Race number: 112. 2nd OA, "Remordu."
        27/4: Knokke. Flying Kilometre. Race number: 111. 2nd OA, Catulle.
        4/5: Cote Dinant. Race number: 102, 3rd OA. Ringoir.
        4/5: Cote Dinant. Race number: 105, 2nd OA. "Remordu".
        18/5: GP de Spa. GT Race, Race number: 32. 6th GT, Ringoir.
        6/7: Cote D'Ardenne. Race number: 0, 1st GT. Ringoir.
        6/7: Cote D'Ardenne. Race number: 0. 2nd OA. "Remordu."
        13/7: Zandvoort GP MG Car club. 1st OA. Race number: 12. "Remordu."
        13/7: Zandvoort GP MG Car Club. Race number: 1, 2nd OA. Ringoir.
        27/7: Trophee D'Auvergne. Race number: 128. DNS. Ringoir.
        17/8: Cote de Bomeree. Race number: 105. DNS. Ringoir.
        17/8: Cote de Bomeree. Race number: 103. DNS. "Remordu."
        14-21/9: Tour de France. DNF (differential failure). Race number: 169.
         Ringoir/Heinzelmann.
        5/10: Cote Namur. DNF (mild frontal damage).
1959:   18-25/9:Tour de France. DNS (withdrawn). Race number: 166. Ringoir.
1959:   Repossessed by Belgian bank.
1962:   Sold for £1200 to Malcolm Bennett and imported dissassembled into UK 1974.
        Sold to Jock Bruce in Guildford, UK.
1974:   Sold to Peter Giddings.
1981:   Restored by Modena Engineering. Registered OPL 199W.
1983:   Sold to Eric Clapton.
1987:   Sold to Prince Zourab Tchkotua. Restored to original configuration by Alligretti of Modena.
1987:   Mille Miglia retrospective. Race number: 278. Tchkotoua/Widdows.
1988:   Mille Miglia retrospective. Race number: 296. David Piper/Robin Widdows.
1993:   Sold to Lance Hill, Canada. Completely restored by August 1996.
1996:   Entered in Pebble Beach Concours.

End of production of the" fourteen louvre" Berlinettas

0723 GT

Delivered to Bjorstrom in Sweden on 6/8/57 for Curt Lincoln of Scuderia Askolin to drive. Three vents on sail panel, covered headlights. *Tipo 508C* chassis. Dark blue with grey top.

1957:   11/8: Swedish GP. 1st in GT. Lincoln.
        18/8: Abu-Turku GT race in Finland. 1st. Lincoln.
        Kristianstad GT race. 1st. Lincoln.
1958:   Helsinki GP in Finland. 1st in GT. Lincoln.
1959:   Midnight Sun Rally. 1st. Race number: 17. Qjarnstrom/Andersson.
        Car severely burnt in Belgium. Rebuilt with one vent and RHD.
1960s: Sold to R. Hurtha/F. Geitel. Displayed in a museum in Helsinki.
1996:   Displayed in the Schlumpf museum.

0731 GT

Delivered to Malle in France on 4/4/57. Scaglietti Comp. Berlinetta. Red with stripe. Registration: GE 66444.

1957:   15-21/9: Tour de France. DNF - crash. Race number: 166. Simon/Jean Aumas.
        27/10: Biere/Marchairuz Hill. 1st in GT. Aumas.
1958:   Sold to Guelfi, Paris.
        7/4: Pau 3 hours. 6th OA. Guelfi.
        13/4: Montlhéry Coupé du Printemps.1st in GT. Guelfi.
        18/5: Spa GP. 3rd OA. Guelfi.

6/7: Rheims 12 hours. DNF. Bad crash. Guelfi/Gurney.
5/10: Montlhéry Coupés du Salon. DNF. Guelfi/Gurney.
1959:   Sold to Cicoira.
    Sold to Thepenier.
    Sold.
1989:   Advertised for sale through Wes Clark Classic Cars, Dallas, Texas.
1996:   Sold to J. F. Malle, France.

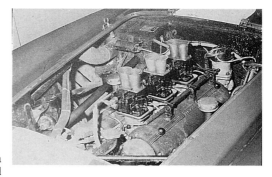

0733 GT

Delivered to Shell of France for Maurice Trintignant on 23/7/57. Sometimes fitted with a Testa Rossa prototype engine and entered by the Scuderia Ferrari. French blue with longitudinal red racing stripe. Scaglietti Comp. Berlinetta with three slots on sail panel and flap on nose giving access to the radiator cap. Registration number: BO 84716.

1957:   26/5: Nürburgring 1000kms. DNF. Race number: 45. Picard/Bruce Kessler.
    16-22/9: Tour de France (Scuderia Ferrari). 2nd OA. Race number: 169. Trintignant/Picard. Bad crash on the night before the event.
1958:   26/1: Buenos Aires 1000kms. 8th OA. Trintignant/Picard.
    Clermont Ferrand. Coupé de Vitesse. 1st in GT. Trintignant.
    27/7: Auvergne 3 hours. 2nd OA. 1st in GT. Race number: 88. Trintignant.
    Sold.
1968:   Sold to Tom Meade, Modena.
    Sold to J. Fisher, USA.
1996:   Still with J. Fisher.

0747 GT

Delivered to Garage Montchoisy for Jean Aumas on 6/9/57. 1081kg. Scaglietti Comp. Berlinetta. Red. *Tipo* 508C/128C. Registration number: GE 54178.

1957:   16-22/9: Tour de France. 3rd OA. Race number: 165. Lucas/Malle.
    6/10: Montlhéry. Coupés de Salon. Lucas/Aumas.
1958:   7/4: Pau 3 hours. 9th OA. Aumas.
    18/5: Dubendorf Slalom. 1st in GT. Race number: 63. Aumas.
    27/7: Auvergne 3 hours. 7th OA, 6th in GT. Race number: 86. Aumas/Schild.
    6/7: Rheims 12 hours. DNF. Race number: 72. Aumas/Wirtz.
    Lelanderon/Lignieres. 2nd in GT. Aumas.
    31/8: Ollon Villars. 4th in GT. Aumas.
    Sold to Jean Pierre Schild, Switzerland.
    14-21/9: Tour de France. 5th OA. Race number: 167. Schild/De la Geneste.
    7/9: La Foucille Hill. 1st in GT. Schild.
    27-28/9: Marchairuz Hill. 1st in GT. Schild.
1960's: Sold to Sherman Fleming in the USA.
    Sold to P. Yevcah.
    Sold to J. Southard.
1978:   Sold to Peter Giddings.
    Sold to M. Colombo in Italy.
    Sold to Eros Crivellari. Now silver with a red stripe.
1982:   Mille Miglia retrospective. Crivellari/Crivellari.
1984:   Mille Miglia retrospective. Crivellari/Crivellari.
1987:   Mille Miglia retrospective. Crivellari/Crivellari.
1988:   Mille Miglia retrospective. Crivellari/Crivellari.
1989:   Mille Miglia retrospective. Crivellari/Crivellari.
1991:   Mille Miglia retrospective. Crivellari/Dell'Isolla.
1996:   Still with Crivellari.

0749 GT

Delivered to Jean Estager in France on 20/7/57. Scaglietti Comp. Berlinetta with three slots on sail panel. 1060kg. Blue with red stripe. 1060kg.

1957 16-22/9: Tour de France. DNF. Crash. Race number: 167. Estager/Schell.

1958:   Sold to N. Da Silva Ramos. Dark green with yellow stripe.
         7/4: Pau 3 hours. DNF. Crash. Race number: 60. Da Silva.
         18/5: Spa GP. 1st in GT. Da Silva.
         6/7: Rheims 12 hours. 7th OA. Da Silva/P. Hill.
         27/7: Auvergne 3 hours. 5th OA. 4th in GT. Race number: 80. Da Silva.
         14-21/9: Tour de France. 3rd OA. Race number: 162. Estager/ Da Silva Ramos.
         Raced by J. C. Vidilles.
1959:   Sold.
1965:   Seen in Paris.
1996:   In private collection for many years.

0753 GT

Delivered to Russel Cowles, an Englishman living in California, on 28/9/57. Scaglietti Comp. Berlinetta with three slots on sail panel. *Tipo 508C/128C/050C.*

1958:   Phoenix Arizona, SCCA. Race number: 44BP. Cowles.
          3 more races in California.
1959:   Sold.
1960's: Sold to L. Knaack.
          Sold to B. Jones, Indiana.
1978:   Sold to M. T. Lynch.
1979:   Sold to P. Giddins.
1985:   Sold to Dr Baumberger, Switzerland.
1987:   Mille Miglia retrospective. Baumberger/Glass.
1989:   Sold to Italy.
          With Swedish bank.

0763 GT

Delivered to Ecurie Francorchamps for Leon Dernier on 21/10/57 in Belgium. Scaglietti Comp. Berlinetta with three slots on sail panel. Red. Tipo 508C/128C. 241bhp at 7000rpm. 1050kg. Registration No: IN 200.

1958:   11/5: Herbeumont Hill. 1st OA. Dernier.
1959:   Nürburgring 1000kms. 10th OA. Race number: 56. Dernier/Bianchi.
         Le Mans Trials. Dernier.
         Upright headlight front end fitted after crash.
1960s: Sold to Schumacher.
         Sold to Menier, D. E. Bioley, Diercks, De Cremer.
1964:   Zandvoort. Benelux Cup. Race number: 130. Schumacher.
1965:   Zandvoort. Benelux Cup. Race number: 86. Schumacher.
         Zolder. Belgian Cup. Schumacher.
1974:   Sold to Jean-Claude Bajol, France.
1986   Mille Miglia retrospective. Bajol/Croce.
1987   Mille Miglia retrospective. Bajol/Durelli.
1988:   Mille Miglia retrospective. Bajol/Durelli.
1989:   Mille Miglia retrospective. Bajol/Durelli.
1990:   Mille Miglia retrospective. Bajol/Phiippsen.
1996:   Still with Jean-Claude Bajol.

0767 GT

Delivered to A. Caraceni in Italy on 13/11/57. Red. Scaglietti Comp. Berlinetta. Three slots on sail panel. *Tipo 508C/128C.* Engine developed 247bhp at 7150rpm according to the data sheet. 8 x 32 axle ratio, 1150kg. Later rear bodywork was altered to look like a SWB, one slot on rear sail panel. Disc brakes fitted. Now rebuilt completely to original specification. Registration number: BA 8 9600. 25651 BA. Roma P16747.

1958:   7/9: Coppa Inter Europa at Monza. 1st in GT? Race number: 75. Caraceni.
1959:   Sold to Vittorio Bernasconi of Bari. Bodywork modified.
1964:   July. Monopoli-Strada Gineste hillclimb. 1st in GT. Bernasconi.
1965:   August. Ascoli-Piceno-Colle San Marco. 2nd in GT. Bernasconi.

1977:   Sold to V. Girolami, Rome.
1980:   Sold to L. Bertolero, Turin . Restored to original bodywork shape.
1993:   Sold to Renzo Beltrami of Alessandria. Restored mechanically. Registered: Roma P16747.
1994:   Mille Miglia retrospective. Beltrami/De Luca.
1995:   Mille Miglia retrospective. Beltrami/Beltrami.
1996:   Challenge Ferrari-Shell. Beltrami.

0771 GT

Delivered to Russel Cowles, Ferrari representative in Hollywood, in USA on 16/12/57. Scaglietti Comp. Berlinetta with three slots on sail panel.

1950s: Sold to H. Schaub, J. Andrews, Dan Polgrew, C. Phelps, John Von Neumann.
1974:   Sold to R. Merrell, USA, who died whilst the car was in restoration.
1988:   Sold to Jim Immke.
1994:   Sold to Junichiro Hiramatsu.

0773 GT

Delivered to Luigi Chinetti in New York, USA on 14/11/57. Scaglietti Comp. Berlinetta with three slots on sail panel.

1957:   8/12: Nassau Speed Week. G. Arents.
1958:   Opa Locka. Orange Bowl Races. D. Mod. Class. 2nd OA. Arents.
         9/3: Boca Raton. C. Prod. 1st in GT. Arents.
         22/3: Sebring. 5th OA, 1st GT. Race number: 22. Paul O'Shea/Bruce Kessler.
         4/5: Danville Race. Georgia. 1st in GT. Arents.
         18/5: Cumberland. 1st in GT. Arents.
         1/6: Bridgehampton. 1st in GT, Class G. Arents.
         8/6: Thompson. C Prod. 1st in GT. Arents.
         15/6: Lime Rock Races. 1st heat, 1st. 6th heat, 2nd OA, 1st in GT. Arents.
1959:   Sold back to Chinetti.
         Sold.
1960s: Sold to H. Hirsh, Cuba.
1960s: Sold back to USA.1974.
         Sold to M. Curley.
1978:   Sold to Pass, Holland.
         Sold to Peter Giddings.
         Sold to M. Colombo, Italy.
         Sold to V. Serventi.
1992:   Sold to Maureaux of France, Registration number: 1859 WW 69.

0781 GT

Delivered to H. Dulles in Switzerland on 20th December 1957. Scaglietti Berlinetta, regular covered headlights, with three slots on sail panel vents. In 1961 converted to Dunlop disc brakes. Never raced. Registration: VD 10634.

1970's: Sold to David Piper, UK.
         Sold to Martin Konig, UK.
1975:   Sold to P. Lindquist, Sweden.
1988:   Mille Miglia retrospective. Lindkvist/Sollevi.
1989:   Mille Miglia retrospective. Lindkvist/Sollevi.
1996:   Still with Peter Lindquist.

0787 GT

Delivered to O. Papais in Italy 15th January 1958. A 1957 Scaglietti. Red. Regular covered headlight Berlinetta with three sail panel vents. Now has engine number 1253 GT.

1958:	16/3: Campionato Universario Modena, 1st GT, Crevellari.
	8/6: Coppa di Asiago, 3rd OA. Papais.
	21-21/6: Mille Miglia Rally, 4th OA, Papais/Crivellari.
1959:	Sold back to Ferrari.
	Sold to D'Orey.
	Monza, Lottery GP. Race number 33, 7th OA. D'Orey.
	Trento Bondone. 2nd GT Class. D'Orey.
	Tour de France. Race number 168, DNS. D'Orey/Drogo.
	Monza, Coppa Inter Europa. Race number 61. DNF. D'Orey/Motor.
1960:	Sold to the USA.
1975:	Sold to J. Boulware, USA.
1980s:	Sold to Pizzolotto, Italy.
1984:	Mille Miglia retrospective. Pizzolotto.
1987:	Mille Miglia retrospective. Marzotto/Pizzolotto.
1996:	With Ferdinand Kroyman.

0793 GT

Delivered to Giovanardi, Italy. Scaglietti, 1957, red, normal covered headlight car. The front of the car received cooling ducts for brakes, an extra wiper on top, and Farina Coupé tail lights in 1959. *Tipo 508C/128C.*

1958:	Serravalle San Mariono, 2nd OA. Giovanardi.
	Bologna San Luca, 3rd GT Class. Giovanardi.
	Pontedecimo-Giovi, 3rd GT Class. Giovanardi.
	Modena Autodromo, Campionato Emilia Romania, 3rd GT Class. Giovanardi.
	Trofeo Montagna, 2nd OA. Giovanardi
	Predappio Rocca della Camine no. 146. 2nd OA, 1st GT, fastest GT. Giovanardi.
	Bolzano-Mendola Hill, 2nd GT Class. Giovanardi.
	Triste-Opicina, 1st GT Class. Giovanardi.
	Stallavena-Boscochiesanuova, 1st in GT Class. Giovanardi.
1959:	Bolzano-Mendola, 6th OA, 1st GT Class. Giovanardi
	Trieste-Opicina, 1st GT and 4th OA. Giovanardi
	Rimini San Marino, 1st GT Class. Race number 182 Giovanardi
	Trofeo della Montagna for 3000cc GT Class. Giovanardi.
	Stallavena-Boscochievanuova, 1st GT Class. Giovanardi.
	Coppa Citta Asiago, 2nd OA, 1st GT Class. Giovanardi.
	Coppa Citta Asiago, 1st GT. Giovanardi.
	Pontedecimo-Giovi, 1st GT. Giovanardi.
1974:	The car was still owned by Mr Giovanardi.
1976:	Sold to Vaccari, Modena, Italy.
1996:	With Mr Vaccari.

0805 GT/0619 GT

January 1958, sold to France, a Scaglietti 1957 Medium Grey. Regular covered headlight Berlinetta with three vents on sail panel. The car was renumbered 0619 GT.
After its racing career the car was sold to an American living in France (an old Ferrari customer). The car was superbly maintained. It was stolen and sustained quite a lot of damage. It was later restored, and sold to a young man who crashed it into the Peugeot factory main wall, badly damaging the car. The engine was removed and inserted in a Farina convertible. Finally the owner of another Berlinetta bought the engine from the Convertible. At the time of writing the car was being restored in the USA.
Registration: 104 EV 30; 619 LP 75; 657 FP 86.

1958:	Auvergne 3 hours, 6th OA, 5th GT. "Pertin" Noblet/Peron.
	Rheims 12 hours, 3rd OA. "Pertin" Noblet/Peron.
	Tour de France No. 159, 15th OA. "Pertin" Noblet/"Lorica" Cavrois.
1959:	Monza Lottery GP, 4th OA. Noblet.

Monza Coppa St. Ambreus, 3rd OA, fastest lap, Noblet.
Trento Bondone Hill. Noblet.
Monza Coppa Europa, 3rd OA. Noblet.
1960:   Montlhéry, Coupés USA, DNF (brakes). Noblet.
        Montlhéry, Coupés Salon. Noblet.
        Sold to Lafond.
1961:   Sold to Don Jetter.
1963:   Sold to Pozzi.
        Sold to Thepenier.
        Sold.
1968:   Sold, wrecked.
1973:   Wreck sold to G. Schmidt.
        Sold to W. Sparling, USA.
1989:   Sold to Fabrizio Violati.
1989:   Mille Miglia retrospective. Thiebat/D'Alpaos.
1990:   Mille Miglia retrospective. Otaki/Sakamoto.
1991:   Mille Miglia retrospective. Otaki/Sakamoto.
1992:   Mille Miglia retrospective. Sandias/Nociti.
1993:   Mille Miglia retrospective. Cannizzaro/Cannizzaro.
1994:   Mille Miglia retrospective. Cannizzaro/Cannizzaro.

Note: Seen in Modena in 1970, at Tom Meade's, a dark blue 1958 Berlinetta with red leatherette upholstery. Three vents on the sail panel, covered headlights and blinker repeaters on the front wing sides (mandatory in Italy from 1960). The car was in excellent shape and bore the chassis and engine number 0805 GT. The plastic headlight covers were bordered by a chrome strip as on the 1962 GTO. Present whereabouts unknown.

0879 GT

Delivered to W. Seidel in Austria in March 1958 (Seidel was a works' driver for Ferrari in 1957). A 1958 Scaglietti covered headlight car, with three vent sail panel (the last to be bodied in this fashion) and sliding windows. Red with black stripe. Perspex "bug deflector" on bonnet. Koni telescopic dampers fitted by factory. Headlights were converted to open style by a body shop (not Scaglietti). Tipo 508C/128C. Registration MO 50823, BO 94477, 44TT. 8 x 34 differential ratio. By 1997 car had been completely restored to original specification.

1958:   7/4: Pau 3 Hours, 2nd OA. Seidel.
        27/4: Trier ATMA, 1st. Seidel.
        18/5: Spa GT, 2nd OA. Seidel.
        15/6: Francorchamps Handicap, 1st. Seidel.
        5-6/7: 12 Heures du Rheims, 4th OA. Seidel/Trips.
        27/7: 3 Heures du Clermont-Ferrand, DNF. Seidel.
        10/8: Karlskoga Kanonloppet, DNF. Seidel.
        15/8: Gaisberg Hill, 1st GT Class. Seidel.
        17/8: Zeltweg, 1st GT Class. Seidel
        7/9: Coppa Inter Europa, DNF. Seidel.
        5/10: Eifelrennen GT, 1st. Seidel.
        12/10: Pfersfeld Airport, 1st. Seidel.
1959:   Wolfsfeld Hill, 1st. Seidel
        Pfersfeld Airport, 1st. Seidel
        Stallavena-Boscochievanuova, 2nd GT Class. Seidel
        Monza Lottery GP, 5th OA. Seidel
        Montlhéry Coupé de Paris, 4th OA, 1st GT Class. Seidel.
        Montlhéry Coupé du Paris, 3rd GT. Ramminger.
        Born Airport, 1st GT Class. Seidel.
        Pfersfeld Airport International race, 1st GT Class. Seidel.
        Rossfeld, 2nd GT. Seidel.
1960s: Sold to Ramminger, Germany.
        Sold to Siegfried Mahnke.
1966:   Sold to G. Schmidt.
1973:   Sold to C. Mellin, Switzerland (Mondial 0498 MD p/exchanged). In storage until 1994. Totally restored by 1995.
1997:   For sale at Brooks auction, Monaco. Sold to someone in Italy.

0881 GT

Delivered 25th February 1958 to Antoine D'Assche, Francorchamps, Van den Bosch. Scaglietti, 1958, normal covered headlight Berlinetta. *Tipo* 508C/128C.

1958:   Took part in some hillclimbs (Cote d'Herbeumont) in Belgium. Burned in a garage fire. Later wreck and spares were sold to M. Bennett.
1962:   Imported into UK by M. Bennett.
1993:   Rebuilt by Stephen Pilkington in UK. Red.

0893 GT

Delivered to G. Reed, USA, on 2nd March 1958. A Scaglietti car, 1958, white with blue stripe, normal covered headlights.

1958:   22/3: Sebring No. 21, 7th OA, 2nd GT Class, Arents/Reed/O'Dell.
        17/5: Cumberland. Race number: 196. Reed.
        22/6: Elkhart Lake, 1st Race, 3rd. Reed.
        16-17/8: Montgomery National, 1st GT Class. Reed.
        ?/12: Nassau Trophy. Reed. SCCA National Champion.
1959:   Sold to A. Pabst.
        Sold to J. Kimberley.
1961:   Sold to Ed Weshler.
1962:   Road America 500, 7th OA. Race number: 78.
        Fred Rediske.
1963:   Road America June Sprints, 14th OA. Race
         number: 78. Bob Birmingham.
1964:   Sold to E. Lake.
1965:   Sold to Eline.
1970's: Sold to Brooks Stevens Museum.
1981:   Sold to A. Woodall.
1982:   Sold to B. Brunkhorst.
1983:   Sold to H. Javetz.
1985:   Sold to J. Marchetti.
        Sold to GP SSR.
        Sold to A. Wang.

0895 GT

Delivered to C. Marchi, Italy, on 21st March 1958. A Scaglietti 1958 regular covered headlight Berlinetta, converted to open lights in 1959. Red. *Tipo* 508C/128C.

1958:   Marchi. Competition history, if any, unknown.
1960:   Sold back to Ferrari.
        Sold to R. W. Porta, USA
1960's: Sold to D. Johnson.
1975:   Monterey historic race, 1st in class.
        Sold to J. Boulware.
        Sold to R. Cowherd.
        Sold to Glen Snider.
1977:   Sold to S. Baumgard.
1983:   Offered for sale by Jon Baumgartner.

0897 GT

Sold to FASF, Italy, on 28th May 1958. A Scaglietti 1958 normal Berlinetta with covered headlights but with one vent sail panel. *Tipo* 508C/128C. The 1958 second version, it has a different engine. Competition history, if any, unknown.

1960s:	Sold to the USA.
	Sold to Earle.
	Sold to R. W. Merrell.
	Sold to Peak.
1977:	Sold to Ziering.
1987:	Sold to Spangler.
1990:	Sold to Stieger.
1995:	Sold to M. Fichts, Germany.
1996:	Mille Miglia retrospective. Fichts.

0899 GT

Delivered to Lualdi Gabardi of Italy in March 1958. A Scaglietti 1958 normal Berlinetta with covered lights and one vent sail panel. In 1959 some vertical holes were added to the area under the grille to assist brake cooling. Registration: VA 489 JN9A

1958:	7/4: Pau 3 hours No 51, DNS. Lualdi.
	1/6: Coppa della Consuma Hill, 1st GT Class. Lualdi.
	15/6: Varese Campo di Fiori, 1st GT Class. Lualdi.
	21-22/6: Mille Miglia Rally. Lualdi.
	6/7: Bolzano/Mandola. Race number: No. 224, 4th OA, 1st GT Class. Lualdi.
	7/9: Monza Coppa Inter Europa, 3rd OA. Lualdi.
	14/9: Lumezzane-Sun Appolonio, 2nd OA, 1st GT Class. Lualdi.
	4/11: Monza Coppa San Ambreus, 1st OA. Lualdi.
	Trofeo Montagna, (Trophy of the Mountains) 1st in GT Class, (over 2600cc). Lualdi.
	Sold to F. Pagliarini.
1959:	Monza Lottery GP , 12th OA. Race number: 39. Pagliarini.
	Castelquarto-Vernasca, 1st OA. Pagliarini.
	Pontedecimo-Giovi, 2nd GT Class. Pagliarini.
	Trieste-Opicina, 7th OA, 4th GT Class. Pagliarini.
	Rimini/San Marino, 3rd GT Class. Pagliarini.
	Stallavena-Boschchievanuova, 3rd GT Class. Pagliarini.
	Consuma Hill, 6th OA, 2nd GT Class. Pagliarini.
	Campiano Vetto d'Enza, DNF (crash). Pagliarini.
1960:	Sold back to Ferrari.
	Sold to Nord Africaine, Algiers, for Mounier Hassi-Messaoud Rally, Sahara Desert, 4th OA. Mounier/Graziani.
	Constantine Hilclimb. Mounier.
	Oran Hillclimb. Mounier.
	Staoueli Circuit, DNF (broken driveshaft). Mounier.
	Rouen GP, 7th OA. Mounier.
	Auvergne 6 hours, 10th OA, 4th GT class. Race number: 23. Mounier/Tavano.
	Left at Modena. At the end of the year the car was wrecked badly and taken to a junk yard in Marseilles.
1970's	Remnants parted out: several people in France own engine, much mangled body and almost-destroyed chassis.
1993:	Car rebuilt.
1993:	Tour de France retrospective. Ferry/Ferry.
1994:	Tour de France retrospective. Ferry/Boyer.
1995:	Tour de France retrospective. Ferry/Boyer.

0901 GT

Delivered to F. Picard, France, on 31st March 1958. A 1958 Scaglietti Berlinetta. French blue with stripe. Lighter brakes, 9 x 34 differential ratio, 600 x 16 wheels. Sometimes used a Testa Rossa prototype engine. *Tipo* 508C/128C. Registration number: BO 98213

1958:	1/6: Nürburgring 1000kms, DNF (crash). Picard.
	27/7: Auvergne 3 hours, DNF (crash). Race number: 78. Picard
	7/4: Pau 3 hours, 5th OA. Picard.
	6/7: Rheims 12 hours, 5th OA. Picard/Burgraf
	14-21/9: Tour de France, Scuderia Los Amigos, 2nd OA. Race number: 160. Trintignant/Picard.
	Car returned to Ferrari and overhauled.

Sold to W. Luftman, USA.

1959:	Sold to D. Bell.
1966:	Sold to A. Simmers.
1976:	Sold to SSR.
	Sold to Bob Grossman.
1977:	Sold to Dr Gilmore.
1978:	Sold to Beverly Hills Motor Cars.
1979:	Sold to Hadjuk.
1980's:	Sold to J. Giels.
1986:	Sold to S. Barney.
1996:	Sold to Erich Traber.

0903 GT

Delivered to T. Bjorstrom, S. Nottorp, Sweden, 21st April 1958. A 1958 Scaglietti Berlinetta. Red.

1958:	6/7: Rheims 12 hours no. 70, DNF. Nottorp/Andersson.
1970s:	Sold to L. Saaf.
1993:	Still in Sweden.

0905 GT

Delivered to Ferraro, Italy, 7th May 1958. A typical Scaglietti 1958 Berlinetta with opening air door on front wings.

1958:	Various races in Italy under the alias "Hippocrate."
1960's:	Sold to the USA.
	Sold to F. Peters.
1970's:	Sold to C. Betz.
1990:	Mille Miglia retrospective. Betz/Betz.
1996:	Still with Chuck Betz.

0907 GT

Delivered to Derstefanian, Italy, 8th May 1958. A normal Scaglietti 1958 Berlinetta. *Tipo* 508C/128C. Body much modified.

1958:	11/5: Targa Florio. Race number: 50. Derstefanian killed during practice.
1975:	Sauro, Italy.
1980:	Sold to E. Ansaloni, restored.
1983:	Sold to Ol Blumenthal.
1986:	Mille Miglia retrospective. Blumenthal-Rossi.
1996:	Still with Mr Blumenthal.

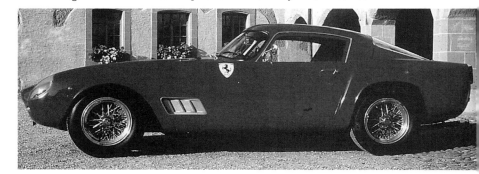

0909 GT

Delivered to Gd. Garage Waberni, W. Lambert, Switzerland, on 14th November 1958. A Scaglietti 1958 normal Berlinetta with one sail panel louvre, covered lights, a chromed strip under the doors. Later this was converted, very poorly, to open lights. Now rebuilt to correct shape. Registration number: SO 5107, TI 32286, AG 35666686

1958:	14/9: Mitholz-Kandersteg Hill, 1st GT Class. Lambert.
	27-28/9: Marchairuez Hill, 2nd GT Class. Lambert.
	15/6: La Landeron Lignieres Hill no 78, 1st GT Class. Lambert.
	31/8: Ollon-Villars Hill, 2nd GT Class. Lambert.
1959:	Slalom de Payern, 1st GT Class. Lambert.
	Mitholz-Kandersteg Hill, 2nd GT Class. Lambert.
	Macon-Solutre Hill no. 86. Lambert.

Truckeim Hill, 1st GT Class. Lambert.
Marchairuz Hill, 1st GT Class. Lambert.
Vaduz-Triesenbert Hill, 1st GT Class. Lambert.
1960: Dubendorf Slalom, 1st GT Class. Lambert.
Payern Slalom, 1st GT Class. Lambert.
Mitholz-Kandersteg Hill, 1st GT Class. Lambert.
Sierre-Montana Hill, 1st GT Class. Lambert.
Ollon-Villars, 1st GT Class. Lambert.
1961: Sold to Lugano.
Sold to Tommy Schpyiger.
Ventoux Hillclimb, 2nd GT Class. Spychiger.
Various other hillclimbs.
1963: Sold to H. Kocher.
1966: Sold to R. Maurer.
1969: Sold to H. Ziegler, Germany.
1974: Sold to R. Sammueller, Germany.
1984: Mille Miglia retrospective. Hofer/Sammueller.
1986: Sold to Kaempfer, Switzerland.
1990: Sold to Juerg Heer of Switzerland.
1994: Sold to Holzhausen of Munich.
1996: Sold to Max Deittel. Displayed at Essen Motor Show. Restoration in red completed.

0911 GT

Delivered to L. Taramazzo, Italy, 28th May 1958. A normal looking 1958 Scaglietti Berlinetta with wind-up windows. *Tipo 508C/128C-0230C*. Red. Registration numbers: IM 19101, 68MM, JMJ 1, TDF 266. At some time the nose was rebuilt slightly off-shape. Completely restored in 1992/3.

1958: 21-22/6: Mille Miglia Rally, 1st OA. Race number 1. Taramazzo/Gerini.
20/7: Garessio San Bernardo, 1st. Taramazzo.
13/7: Trento Bondone, 1st GT Class. Taramazzo.
7/9: Monza Coppa Inter-Europa, 1st OA. Taramazzo.
28/9: Pontedecimo-Giovi, 2nd OA, 1st GT Class.
4/11: Monza Coppa San Ambreus, 2nd OA. Taramazzo.
1st in Italian GT Championship over 2600cc.
1959: Monza Coppa San Ambreus, 5th OA. Taramazzo
Sold back to Ferrari.
Sold to G. Gazzoni Frascara.
1960s: Sold to UK.
Sold to John Broad.
Sold to A. Stamer.
1974: Sold to H. Cluxton, USA.
1977: Sold to Rex K. De George, USA.
1978: Sold to Ed Niles, USA.
1979: Sold to R. Taylor.
Sold to D. Autrey.
1980: Sold to John Starkey, UK.
1982/86: Mille Miglia retrospective.
1984: Zandvoort. FOC Race. 7th OA. Starkey.
1984: Mille Miglia retrospective.
1985: Zandvoort. FOC Race. 8th OA. Starkey.
1986: Mille Miglia retrospective.
1986: Oulton Park. FOC Race. 10th OA. Starkey.
1988: 10/8. Nürburgring "Old Timer" meeting. 1st OA. Starkey.
1989: Zandvoort FOC Race. Starkey.
1991: Silverstone. '50s Sportscars. 12th OA. Starkey.
1992: Sold to Reza Rashidian, UK. Completely restored by D.K. Engineering and RS Panels.
1995: Zandvoort. FOC race.

0925 GT

Delivered 26th July 1958 to Ferrari representative Hollywood, USA, O. Zipper. A normal Scaglietti, 1958 Berlinetta. *Tipo 508C/128C*.

1958:	Sold to Harrah.
1986:	Sold.
1989:	Sold to Baron Mayr-Mellenhof of Austria.
1992:	Sold to Kato in Japan.
1993:	Sold to Harald and Inge Mergard, Germany.
1993:	Tour de France. Mergard/Mergard.
1993:	Mille Miglia retrospective. 221st. Hoyle/Boyce.
1993:	Coppa d'Oro. DNF. Race number: 84. Hoyle/Boyce.
1993:	Tour de France retrospective. Mergard/Mergard.
1994:	Mille Miglia retrospective. Race number: 16. Mergard/Mergard.
1994:	Coppa d'Oro. Mergard/Mergard.
1995:	Restored.
1995:	Mille Miglia retrospective. Mergard/Mergard.

0931 GT

Delivered to Ghersi, Italy, 26th May 1958. A normal Scaglietti 1958 Berlinetta with covered lights. Later the rear end was modified in order to look like a SWB: the rear wings were shaped exactly as a 1960 car.

1959:	Sold to Switzerland.
	Sold to M. Saugey (kept for 10 years).
1970:	Sold.
1975:	Sold to H. Savoy, France.
1976:	Badly wrecked.
1977:	The wreck was sold in Belgium and the car restored retaining the body modifications. Edmund Pery

0933 GT

Delivered to C. Toselli, Italy, 17th June 1958. A normal Scaglietti 1958 Berlinetta. Gold. Wind-up windows, leatherette interior. In 1960 used air filters. In 1960 the engine was upgraded with some very light con rods (432 grams, higher pistons and a magnesium TR sump was also fitted). In 1961 the car was abandoned by its French owner at Paris Airport and impounded by customs. Stored for seven years, then auctioned. Registration: ND 0943, 2910 PJ 92.

1958:	19/10: Aosta-Pila, 1st GT Class. Toselli.
	Sold to Fayen, France.
1959:	Monza Lottery GP, DNF. Race number: 41. Fayen.
	Exported to Venezuela and raced there by Fayen.
1960:	Used in France; abandoned.
1967:	Sold to D'Epenoux.
1968:	Sold to Bardinon.
1975:	Sold to F. Chandon.
1977:	Sold to Col. Mas du Clos.
1983:	Sold to Dominuque Bardini. Mechanical components overhauled. Car still has less than 40,000kms since new.
1985:	Sold to Seydoux.
1989:	Sold to Ogliastro.
1995:	Unstamped engine, sold to Andre Binda, Mougins.
	Sold to Schulte.
1996:	For sale in Germany (DM685,000). Now has "0933 GT" stamped on engine. Red with original red/brown interior.

0967 GT

Delivered to Ferrari representative, Hollywood, USA, 31st July 1958. A normal Scaglietti Berlinetta with covered lights and one vent on sail panel. *Tipo* 508C/128C. Delivered with a chrome roll bar and a ribbed gearbox.

1958/9:	Several races in California with Richie Ginther.
	Perhaps some races by R. Bucknum and then Forbes.
	Robinson in Hawaii.
	1/8: Sold to Stanley.
1960:	Sold to M. Jensen.
1960s:	Sold to General Motors, Detroit, for study.

All mechanical components rebuilt by GM.
Sold to Warren Fitzgerald.

1977:   Sold to R. Bodin.
1979:   Sold to J. Pourret.
1979:   Sold to H. Chambon, France.
1985:   Sold to Heinrich Kaempfer, Switzerland.
1988:   August. Entered "Oldtimer GP" Nürburgring. Kempfer.
1991:   Sold to Yoshiho Matsuda.
1994:   Mille Miglia retrospective. Matsuda/Matsuda.

0969 GT

Delivered W. Mairesse, Belgium, 4th July 1958. A normal Scaglietti 1958 Berlinetta with wind-up windows. White, 1100kg (2428lb). 600 x 16 wheels. 9 x 34 and 8 x 34 differential ratios. *Tipo 508D/128C-02.8C*. As 0969GT was owned by Mairesse and was frequently damaged on the front left wing! Registration numbers: BO 21028, BO 06512, BO 14675

1958:   6/7: Rheims 12 hours, 2nd OA. Mairesse/Beurlys, EF.
         27/7: Auvergne 3 hours, 3rd OA, 2nd GT class. Race number: 72 Mairesse.
         14-21/9: Tour de France, DNF. Race number: 168. Mairesse/Desse.
         Montlhéry 1000kms, 2nd OA. Mairesse/Von Trips.
1959:   Monza Lottery GP, 3rd OA. Race number: 24. Mairesse.
         Lyon-Charbonniere, DNF (bad shunt left rear). Race number: 1. Mairesse.
         Tour de France, 2nd OA, 2nd GT Class, 3rd Index. Race number: 159. Mairesse/Berger.
1960:   Traded in at Ferrari.
         Sold to R. Porta, USA.
1962:   Sold to R. Hattaway.
1960s: Sold to Rezzaghi Motors.
1965:   Sold to Larry's Foreign Car Service - badly wrecked.
         Sold to C. Meideros.
1968:   Sold to J. Boyd.
1987:   In USA.
1996:   Still with J. Boyd.

0971 GT

Delivered to P. Fuare, Switzerland, 9th September 1958. A normal Scaglietti 1958 Berlinetta. 9 x 34 differential ratio. 1127kg (2487lb), wind-up windows. *Tipo 508D/128C*. In 1960 the car was converted to open headlights by Scaglietti after a crash. Registration number: BO06099.

1958:   14-21/9: Tour de France, 8th OA. Race number: 166. Faure/Paulard.
1959:   Picardie Rally, 1st OA. Faure/Paulard.
         Touquet Rally, 1st. Faure/Faure.
1960:   Sold back to Ferrari.
         Sold to Ecurie Francorchamps.
         Montlhéry 1000kms, 9th OA. Race number: 15. Pilette/Berger.
1960s: :Sold to L. Sjostrom, Sweden.
1973:   Sold to Ove Haak - red.
1990:   Sold to Sollavi. Rebuilt to covered headlights & converted to discs & telescopic dampers.
1993:   Sold to Baron Hatzfeld (Black). At this time, the car had only 30,000 miles.
1993:   Tour de France retrospective. Hatzfeld/Wild.
1994:   Tour de France retrospective. Wildenburg/Hahne.
1996:   With Hein Gericke.

0973 GT

Delivered to Bourillot, France, 30th August 1958. A normal Scaglietti 1958 Berlinetta. Leather upholstery, wind-up windows. Ferrari red (*not* racing red). 1100kg (2428lb). In 1971 the engine was removed and 0949 GT installed. Original engine is in 0849 GT. Registration numbers: BO 06502, 0047414, 46 CZ 53

1958:   14-21/9: Tour de France, 7th OA. Race number: 165. Bourillot/Masetti
        5/10: Montlhéry Coupés du Salon. Race number: 3. Bourillot.
1959:   Montlhéry Paris GP, 2nd GT Class. Bourillot
        Montlhéry Coupés USA, 1st GT Class, Bourillot.
        Sold to A. Simon.
        Tour de France, DNF (sump broken by stones near finish). Race number: 173. Simon/Thepenier.
1960:   Montlhéry Paris GP, winner. Race number: 22. Simon.
        Montlhéry Coupés USA. Simon.
        Limousin Rally, 6th OA, 2nd GT Class. Simon/Barthe.
        Bordeaux Sud Ouest Rally, 6th OA. Simon.
1961:   Sold to Loyer.
        Sold to Houdusse.
1965:   Sold to Pourret.
1967:   Sold to Vernholes.
1971:   Engine replaced with that from 0949 GT (a coupé). 0973 GT engine installed in 0849 GT.
1972:   Sold to Thuysbaert.
        Sold to Prost.

1031 GT

Delivered to J. Peron, France, 12th September 1958. A normal Scaglietti 1958 Berlinetta, wind-up windows. Blue. 1100kg (2428lb). 600 x 16 wheels. *Tipo* 508D/128C. Registration number: BO 06520.

1958:   14-21/9: Tour de France, 4th OA. Race number: 163. (shunt front left). Peron/H. Shell.
1959:   Sold to B. Cotton, Paris.
        Monza Lottery GP. 6th OA. Cotton.
        Tour de France, DNF. Race number: 162. Cotton/Beudin.
        Montlhéry Coupés Paris, DNF. Beudin.
1960:   Sold.
1960s: Sold to D. Love, USA.
1991:   Colorado Grand. Love/Love.
1992:   Colorado Grand. Love/Love.
1993:   Colorado Grand. Love/Love.
        Sold to Mary Hoe-Love.
1994:   Behring Auto Museum, Danville, USA.
1995:   Colorado Grand. Love/Love.
1996:   Still with Love.

1033 GT/1523 GT

Delivered to Gendebien, Belgium, 12th September 1958. A normal Scaglietti 1958 covered headlight Berlinetta. Metal grey. Sliding windows, oil radiator, twin master cylinders. *Tipo* 508D/128D-08D. Very fast engine. 252bhp at 7200rpm. 1000 kg. Following crash in 1959 the front end received open headlights at Scaglietti and the car was renumbered 1523 GT. At this time, Bianchi owned both cars. In 1960 Campagnolo disc brakes were installed. Registration numbers: BO 06532, 4289 MP78.

1958:   14-21/9: Tour de France, winner. Race number: 164.
        Gendebien/Bianchi.
        5/10: Montlhéry Coupés du Salon, 1st OA.
        Gendebien/Bianchi.
1959:   Montlhéry Paris GP, 1st GT Class. Gendebien.
        Nürburgring 1000kms, 10th OA, 2nd GT Class.
        Bianchi.
        Monza Lottery GP, 11th OA. Race number: 27. (Bad crash). Bianchi.
        Rebuilt at factory with open headlight nose.
        Sold to Marinelli in Switzerland.
        Sold to Brunetti, Switzerland.
        Sold to Wildhabert, Caleche, Switzerland.

1967:   Sold to De Siebenthal.
1968:   Sold to Debord, Paris.
1970:   Sold to Badre, Paris. When Badre bought the car, he wrote to the author, pointing out that it had been re-numbered to 1523 GT by the factory in 1959.
1982:   Sold to Baily in France.

1035 GT

Delivered to Gomez-Mena, Cuba, 23rd February 1959. A normal Scaglietti 1958 Berlinetta. Grey. 1100kg (2428lb). Wind-up windows, chrome embellishers around headlights. 600 x 16 wheels. *Tipo 508D/128C-086C.* Registration numbers: BO 06533, 250 TDF.

1958:   14-21/9: Tour de France, DNF. Race number: 161. Gomez-Mena/Meyer.
1959:   Ojeda City GP, (Venezuela). Gomez-Mena.
1960:   Cuba GP, 10th OA, 2nd GT Class. Gomez-Mena.
1975:   USA, W. Richmond.
1977:   Sold to P. Giddins.
        Sold to D. Margulies, UK.
        Sold to B. Classic.
1978:   Sold to M. Hilton.
1983:   Sold to M. Spitzley.
1992:   Tour de France retrospective. Spitzley/Spitzley.
1994:   "Ecosse" tour of Scotland. Spitzley/Spitzley.

1037 GT

Delivered Kaufman, Chimeri, Venezuela, 21st November 1958. A normal Scaglietti 1958 Berlinetta. White. Wind-up windows.
1959:   Trinidad GP, 3rd OA, 1st GT Class. Chimeri.
        Ojeda City GP. Chimeri.
1960:   Cuba GP, DNF. Chimeri killed when his 250 Testa Rossa left the track, hit a ravine and exploded. After the crash the Berlinetta was sent back to Scaglietti and rebodied with a 1959-style body
        with open lights. Like the other rebodies (0619/0805) fitted with California LWB wings which used moulded wheel openings. Front bumber with upright bumperettes also fitted.
1980s: Sold to Joe Hirth.
1988:   Sold to Mike Sheehan.
        Sold to John Maher, Australia. Damaged in transit; restored by Auto restorations of New Zealand.
1991:   Sold to Jack Braam Ruben in Germany.
1992:   Sold to Peter Fandel of Bitburg, Germany.

1039 GT

Delivered to Chinetti/H. Harcourt, USA, 26th November 1958. A normal Scaglietti 1958 Berlinetta. *Tipo 508D/128D.*

1959:   Sold to H. G. Peters (may have raced at Sebring).
1964:   Sold to Greenwich Autos. Engine and transmission replaced by Ford units.
1974:   Sold to C. Betz. Restored with original engine and gearbox.
1980s: Sold to L. Menser.
1984:   Monterey Historic Races. L. Menser.
1987:   June. Sold to Joe Moch, Grand Rapids.
1994:   Sold to Jean-Pierre Slavic, Switzerland. Registered: VD 433 U.

1113 GT

Delivered to W. Sturgis, USA, 13th November 1958. A normal Scaglietti 1958 Berlinetta.

1959:   Pomona *Los Angeles Examiner* GP. Sturgis.
        Laguna Seca, Class EM, 1st. Sturgis.
        Elkhart Lake 500, Class DM, 1st. Sturgis/Bondurant.
        Sold to Hollywood Sports Cars.
        Sold to A. Mourabian.
1960s: Sold to J. Gaughan.
1996:   Still with John Gaughan.

1127 GT

Delivered to G. Reed, USA, 13th November 1958. A normal Scaglietti 1958 Berlinetta, roll bar, cylinder heads milled 25/100mm, very light con rods. Maroon. *Tipo* 508D/128D-040D.

1959:	Sebring, 3rd GT class, Reed/Arents/Arents/O'Dell.
	1959 SCCA GT Champion.
1960s:	Sold to J. McNeil.
	Sold to M. Green.
	Sold.
1973:	P. Pappalardo.
1978:	Sold to Resnick.
1980:	Sold to A. Woodall.
1986:	Sold to Len Rusiewicz. PA.

1139 GT

Delivered to Ferrari representative Hollywood, USA, 18th December 1958. A normal Scaglietti 1958 Berlinetta. *Tipo* 508D/128D-120D. Now has roll bar.

1960s:	Sold to R. Von de Water.
1978:	Sold to J. Moch.
	Sold to J. Marchetti.
1980:	Sold to E. Weschler.

1141 GT

Delivered to Luigi Chinetti and George Arents in USA on 30/12/58. Scaglietti Comp. Berlinetta. Silver.

1961:	Sold to J. Mecom.
1960s:	Sold to Jim Hall.
1970s:	Sold in Colorado.
1990:	Engine, minus starter, for sale.
1994:	Sold to Peter Glaesel of Germany. Silver.
1994:	Tour de France retrospective. Glaesel/Saft.
1996:	For sale.

1143 GT

Delivered to SOCONEMET SA for Jean Aumas in Switzerland on 11/2/59. Red with white stripe. Scaglietti Comp. Berlinetta with open headlights. Registration numbers: GE 54178, BS 14194, GL 2390, ZH 89457, ZH 40066, SG 18862, LU 9918.

1959:	Mille Miglia Rally, 3rd OA, Race number: 10. Schild/Pecorini.
	Eau Morte Kilometre, 1st OA, 1st in GT. Schild.
	Verbois Hill, 1st in GT. Schild.
	Iseran/Mt. Mlanc Rally, 1st OA. Aumas/Schild.
	Lyons/Charbonniere Rally, 4th OA. Race number: 2. Schild.
	Lottery GP at Monza, 10th OA. Race number: 9. Schild.
	Mitholz-Kandersteg Hillclimb, 1st in GT. Race number: 107. Schild.
	Sold to A. Hopf, Switzerland.
	Tour de France. DNS. Race number: 172. Berney/Gretener.
1961:	Sold to F. Hatchec of Switzerland.
1962:	Sold to R. Hauser.
1964:	Sold to E. Frischknecht.
1960s	Sold to Mosberger.
	Sold to H. Vallaster.
1970s:	Sold to G. Frey. Registration number: ZH 102 792.
1982:	Mille Miglia retrospective. Frey/Frey.
1992:	Tour de France retorospective. Frey/Aebi.
1993:	Tour de France retorospective. Frey/Aebi.

1161 GT

Delivered to L. Chinetti in USA on 11/3/59. Scaglietti Comp. Berlinetta. British Racing Green.

1959:	Grossman?

Sold to P. Sherman.
1960s: Sold to Delameter.
Sold to K. Hutchinson.
1995: For sale in the UK.

1959 Berlinettas, 1st version. One vent on sail panel. Most cars with open headlights, but not all. *Tipo* 128D engines

1309 GT

Delivered to De Micheli in Italy on 24/3/59. The first 1959 car with open headlights. *Tipo* 508D/128D-0278D. Presently has engine from 3917 GT and covered headlights. Registration number: F1 116892. Red and black.

1959: Bolzano-Mendola, 3rd OA. De Micheli.
Castelfusano Flying Kilometre, FTD. Race number: 112. De Micheli.
Coppa Borzachini-Terni, 1st. De Micheli.
Predappio Rocca della Carmine Hillclimb, 3rd OA,
1st in GT. De Micheli.
Strettura Passo Della Somma, 1st. De Micheli.
Trofeo Bettoja. 2nd OA, De Micheli.
Coppa Fagioli, Castelfusano, 3rd OA. De Micheli.
Stallavena-Boscochievanuova, 4th in GT. De Micheli.
Rimini-San Marco, 4th in GT. De Micheli.
Bolzano-Mendola, 3rd in GT. De Micheli.
Coppa Citta Asiago., 3rd in GT. De Micheli.
Sold to Marie-Gabrielle de Savoie, Switzerland.
1960s: Sold to G.M. Focquet in UK.
Sold to R. Cowles.
1970s: Sold to A. G. Livic.
Sold to C. F. Weber, USA.
1984: The author travelled in the car whilst visiting Laguna
Seca. It was then in very good condition with the engine
from 3917 GT (GTE) installed. Chuck Weber raced the
car at that year's Monterey historic races.
1984: Correct engine for sale in Switzerland.
1992: Correct engine installed.
1992: Sold at Brooks auction.
1993: Sold by the Chequered Flag in USA.
Sold by Forza.
Sold to Gary Roberts.
Sold to Bill Jacobs.
1994: For sale at Talacrest, UK.
1995: Sold.

1321 GT

Delivered to Garage Francorchamps for Jan Blaton (Beurlys) in Belgium on 27/3/59. Open headlight car. Red with yellow stripe. *Tipo* 508D/128D-0280D. **252bhp at 7200rpm. 1000kg.** Did have engine number 3477 GT fitted, but now has original engine rebuilt. Registration number: 10 VZ 6.

1959: Laroche Hill, 1st. Blaton.
Bomeree Hill, 1st. Blaton.
Nürburgring 1000kms, 9th OA, 1st in GT. Race number: 35. Blaton/Blaton.
Le Mans, 3rd OA, 1st in GT. Race number: 11. Blaton/Dernier.
1960: Sold to Ramminger, Germany.
1967: Sold to H. Diesperger in Germany.
Sold to M. R. Lampe.
Sold to B. Hughes.
Sold back to M. Lampe, USA.
1982: Sold to B. Brunkhorst.
1984: Sold to Robert Des Mardis, Canada.

1333 GT

Delivered to Carlo Mario Abate in Italy on 9/4/59. Red. Scaglietti Comp. Berlinetta.

1959:   Coppa Consuma hillclimb, 5th OA, 1st in GT. Abate.
           Trento-Bondone, 1st in GT. Abate.
           Mille Miglia Rally, 1st OA. Abate/Balzarini.
           Rimini-San Marino, 2nd in GT. Abate.
           Coppa San Ambreus, 2nd in GT. Abate.
           Garessio-San Bernardo, 3rd OA, 3rd in GT. Abate.
           Lottery GP Monza, 2nd OA. Abate.
           Trieste Opicina, 5th OA. 2nd in GT. Abate.
           Coppa Inter-Europa Monza, 2nd OA. Abate.
           Corsa dei Colli Torinesi, 1st in GT. Abate.
           Tour de France, 5th OA, 5th in GT. Race number: 170. Abate/Balzarini.
1960:   Nürburgring 1000km, 8th OA, 1st in GT. Race number: 77. Abate/Davis.
           Ventoux Hillclimb, 5th OA, 1st in GT. Abate.
           Garessio Barnardo, 1st in GT. Abate.
1960s: Sold to Newens.
           Sold to Brookbank.
           Sold to David Black, UK. After David Black's death, in 1990, car inherited by his daughter, Nicki Jackson.
1997:   Sold to Stephen Pilkington.

1335 GT

Delivered to Carlo Toselli of Italy on 23/4/59. Red. Scaglietti Comp. Berlinetta with open headlights. Registration number: TO 281660.

1959:   Bolzano-Mendola, 2nd in GT. Toselli.
           Trieste-Opicina, 6th OA, 3rd in GT. Toselli.
           Lottery GP at Monza, DNF. Race number: 7. Toselli.
           Giro di due Mare, 6th OA. Toselli.
           Garessio San Bernardo, 2nd OA, 2nd in GT. Toselli.
           Coppa San Ambreus at Monza, 2nd in GT. Race number: 187. Toselli.
           Coppa Inter Europa Monza, 3rd OA. Toselli.
1960:   Garessio San Bernardo, 2nd OA, 1st in GT. Toselli.
           Coppa Dell Consuma Hillclimb, 1st. Race number: 236. Toselli.
           Circuito di Posilipo, 1st in GT. Toselli.
           Stallavena-Boscochiesanuova, 2nd in GT. Toselli.
           Sold to R. Bruno in Italy.
1972:   Rear end rebuilt incorrectly.
1974:   Sold.
1974:   Sold to G. Litrico.
           Sold to E. Buzzi.
           Sold to B. Riccardi.
1980:   Sold to Albert Obrist, Italy.
1980:   Sold to G. Schoen, Italy. Registration number: TO 281660.
1985:   Restored to original condition.
1988:   Mille Miglia retrospective. Schoen/Ceresi.
1989:   Mille Miglia retrospective. Schoen/Ceresi.

1353 GT

Delivered to P. Ferraro in Italy on 11/5/59. Scaglietti Comp. Berlinetta with covered headlights. *Tipo* 508D/128D. Black. Now yellow with white stripe. Now has different 250GT engine fitted but the restored original engine is with the car's owner.

1961:   Sold.
1960s: Sold to H. Schmidt, USA.
1985:   Sold.
1991:   Sold to Peter Hannen, UK.
1994:   Sold to Andrew Pisker, UK. Yellow with white stripe.
1996:   Sold.

1357 GT

Delivered to Pierre Dumay of France on 23/4/59. Scaglietti Comp. Berlinetta with open headlights. *Tipo* 508D/128D-0316D. Registration numbers: BO 16636, VA 70731, TN 34854.

1959:	Bouzarea GP, 1st. Dumay.
	Route du Petrole GP, 1st. Dumay.
	Philippeville-Hassi Messaoud-Constantine Rally, 1st. Dumay/Briet.
	Urcy Hillclimb, 6th OA. Dumay.
	Faucille Hillclimb, 1st in GT. Dumay.
	Circuit Vitesse Ataoueli, 1st. Tavano.
	Murojadjo Hillclimb, 1st. Dumay.
	La Grenouillere Circuit, 1st. Dumay.
	Cote du Pin Hillclimb, 1st. Dumay.
1960:	Sold to Lualdi Gabardi, Italy.
	Bolzano-Mendola, 1st in GT. Lualdi.
	Varese Campo di Fiori, 1st in GT. Race number: 128. Lualdi.
	Salsomaggiore San Antonio, 1st in GT. Race number: 228. Lualdi.
	Targa Florio, 10th OA, 1st in GT. Lualdi/Scarlatti.
	Giro di Posilipo, 2nd OA. Race number: 84. Lualdi.
	Stallavena Boscochiesanuova, 1st in GT. Race number: 592. Lualdi.
1960:	Sold to Zampero.
	Coppa Inter Europa at Monza. Zampero.
	Coppa d'Oro at Modena. Zampero.
	Coppa Nevegal, 1st OA. Zampero.
1965:	Sold to P. Civati, USA.
1960s:	Sold to J. Euhrman.
	Sold to I. Cabraloff.
1973:	Sold to H. Cover.
1978:	Sold to Karl Dedolph, Minnesota, and totally restored. Red.
1991:	Sold to J. W. Marriot, Jr., USA. Registration number: CUF 250.

1367 GT

Delivered to SILE for Vladimiro Galluzi in Italy on 26/7/59. *Tipo* 508D/128D. White. The last Zagato-bodied Berlinetta. No "double bubble" roof, but covered headlights and 136 litre fuel tank. Very fast *Tipo* 128D engine with 9.4:1 Compression ratio.

1960:	Sold to Cornelia Vassali (Camillo Luglio's wife)
1975:	Sold to J. Boulware.
	Sold to R. Gatien, USA.
1977:	Crashed.
1978:	Sold to J. Owen.
1980:	Sold to Brigato, Italy. Registration number: PD 748456.

1377 GT

Delivered to Carlos Kaufman for Max Meyer of Venezuela on 10/6/59. The 1959 second version body prototype by Pininfarina. 128 DF engine with outside plugs. Sump and camshaft covers cast in Electron. 40DCZ Webers. 9.5:1 compression ratio. Registration number: EE 8065.

1959:	Le Mans 24 hours. 6th OA, Race number: 20. Fayen/Munaron.
	Shipped to Venezuela.
	Ojeda City GP. Fayen.
1960:	Buenos Aires 1000kms, 9th OA, 1st in GT. Munaron/Todaro.
	Returned to Maranello for overhaul.
	Sold to L. Chinetti in August. Some races by George Arents.
1960s:	Badly crashed and sold to Cherubini.
1967:	Engine removed.
1977:	Car at W. Sparlings.
1996:	Now returned to Luigi Chinetti. Restored.

1385 GT

Delivered to L. Piotti of Italy on 2/5/59. Scaglietti Comp. Berlinetta with open headlights. *Tipo* 508D/128D-0310D.

1959:	Lottery GP Monza, DNF. Race number: 22. Piotti.
1960:	May be car of Fiaccadori.
1960s:	Sold to Stettelini of Monaco. Fitted with Amoradi disc brakes and Koni telescopic dampers by the factory.
1975:	Sold to Pedretti.
	Sold to W. Luftman.
	Sold to J. Morgan, USA.
1979:	Sold to Gelles.
1980:	Sold to Peter Giddins.
1981:	Sold to Gary Schoenwald.
1982:	Sold to Ernie Mendicki.
1994:	Sold to Mike Sheehan.
1996:	For sale, 30,000kms, discs, konis. Red/black.
1996:	Sold to Phillipe Marcq.

1389 GT

Delivered to Rotaflex for Alfonso Thiele of Italy on 29/5/59. Red. Scaglietti Comp. Berlinetta with open headlights. Registration number: NO 52388.

1959:	Coppa San Ambreus Monza, 4th in GT. Thiele.
	Lottery GP Monza, 1st OA. Thiele.
	Coppa Inter-Europa Monza, 1st OA. Race number: 77. Thiele.
1960:	Sold to someone in France.
1960:	Sold to F. Berson, France.
1970:	Sold to F. Lechere.
1972:	Sold to J. Guichet.
1981:	Sold to Schoenwald.

1399 GT

Delivered on 22/5/59 to FBA in Paris, France. Metal grey. Scaglietti Comp. Berlinetta with covered headlights. Never raced.

	Sold to C. Marquant in France.
1996:	Still in France.

1401 GT

Delivered to Luigi Taramazzo in Italy on 23/5/59. Scaglietti Comp. Berlinetta with open headlights. 1133kg. *Tipo* 508D/128D-0376D. Registration number: GE 34675.

1959:	Garessio San Bernardo, 1st OA. Taramazzo.
	Lottery GP Monza, 9th OA. Race number: 6. Taramazzo.
	Mille Miglia Rally, DNF, crash.
	Sold to Spinedi of Switzerland. Painted Gold.
	Marchairuz Hill, 2nd in GT. Spinedi.
	Vaduz-Triesenberg, 2nd in GT. Spinedi.
1960:	Geneva Rally, 1st OA. Spinedi.
	Tour de France. Race number: 158. Spinedi.
	Coupé des Alpes. Race number: 68. Spinedi/Briffault.
	Auvergne 6 hours, 9th OA. Race number: 24. Spinedi/Schiller
	Sold.
1960s:	Sold to Rob de la Rive Box.
	Sold to K. Hasler.
	Sold to P. Hass.
1980s:	Engine block was used in the build-up of Testa Rossa 1720 GT owned by Paul Schouwenburg.
1996:	Still with Plino Haas.

1959 Berlinettas, "Interim", 2nd series body style. 128d/DF engines

1461 GT

Delivered to Luigi Chinetti in the USA on 17/6/59. NART Team car. Scaglietti Comp. Berlinetta Yellow. Registration number: 00483 L4.

1959:   Le Mans 24 hours, 4th OA. Race number: 18. Pilette/Arents.
        Tour de France, DNF, bad crash. Race number: 164. J. Lucas/J.F. Malle.
        New car built at factory using parts from the wreck and fitted with disc brakes and Koni telescopic dampers. Testa Rossa engine No: 0770
          TR fitted.
1962:   Sold to J. F. Malle, Switzerland.
        Sold to Harry Theodoracopoulos, USA.
1970s: Sold to D. Mayberry.
1978:   Sold.
1979:   Sold to M. Colombo, Italy.
        Sold to Cavallino Collection.
1991:   Sold to Fabrizio Violati, "Maranello Rosso."

1465 GT

Delivered to U. Satori in Italy on 2/9/59 with inside plug engine. *Tipo 508D/128D*. Amadori disc brakes fitted in 1960.

1964/5: Raced in Italy by C. Pietro Marchi.
1965:   Imported by Hans Tanner, USA.
1969:   Sold to J. Vulpe.
1976:   Sold to Stan Nowak.
        Sold to Dan Margulies, U.K.
1977:   Sold to M. Colombo.
        Sold to V. Fachini.
1986:   Sold to Classic Car Associates, Holland.

1509 GT

Delivered to Jo Schlesser of France on 8/9/59. Scaglietti Comp. Berlinetta. White with green stripe. *Tipo 508D*. Fitted with inside plug 128DF-0398DF engine. Now has outside plug engine. Registration number: MO 57200,00 459 L 4, VD 2616, 250 EJ 92.

1959:   Tour de France, DNF, driven into river. Race number: 158. Sclesser/Schlesser.
1960:   Nürburgring 1000kms, 11th OA, 2nd in GT. Schlesser/Bianchi.
        Rouen GP, 2nd OA, 1st in GT. Schlesser.
        Auvergne 6 hours, 21st OA. Race number: 80. Schlesser.
        GT race at German GP Nürburgring, 1st in GT. Schlesser.
        Tourist Trophy, Goodwood, 7th OA. Race number: 5. Schlesser
        Sold to Berney, Switzerland.
        Tour de France, DNF. Race number: 160. Berney/Gretener.
1961:   Sold in Switzerland.
1964:   Rally Lorraine. Miro/Courtois.
1960s: Sold to Gerber.
1974:   Sold to Bernard Consten, France. Completely restored.

1519 GT

Delivered to SOCONOMET for J. P Schild of Switzerland on 12/9/59. Red. Scaglietti Comp. Berlinetta. *Tipo 508D/128DF-0478DF*. First delivered with inside plug engine, later fitted with outside plug engine. Registration: GE 54178.

1959:   Tour de France, 3rd OA, 3rd in GT. Race number: 161. Schild/De la Geneste.
        Montlhéry 1000kms. Schild.
        Coupés du Salon, Montlhéry, 1st in GT. Race number: 3. Schild.
1960s: Exported to the USA.
        Sold to J. Bunch.
        Sold to J. Doll.
        Sold to Bob Grossman. Crashed, engine removed.
        Sold to Degner, New York.

| 1976: | Sold to J. Damron, Illinois without engine. (the engine is in 2237 GT SWB). |
| 1996: | Mr Damron is considering restoration. |

1521 GT

Delivered to Pierre Dumay of France on 16/9/59. Metal Grey. Scaglietti Comp. Berlinetta. 40DCL Weber carburettors fitted. *Tipo* 508D/128D inside plug engine taken out in 1969 and an outside plug engine fitted. *Tipo* 128DF-0480DF. Registration numbers: 01086 L 4, 00484 L 4.

1959:	Tour de France, 7th OA, 7th in GT. Race number: 171. Dumay/Daboussi.
	Circuit de Vitesse de Staoueli, 2nd OA. Dumay.
	Staoueli GP, 2nd OA. Dumay.
	Constantine Hillclimb, 1st OA. Dumay.
1960:	Alger/Hassi Messaoud Rally, 1st OA, 1st in GT. Dumay.
	Campagnolo disc brakes fitted in Italy.
	Rouen GP, 4th in GT. Dumay.
	Coupés USA, Montlhéry, 4th OA, 1st in GT. Dumay/Tavano.
	Paris GP, Montlhéry, 3rd OA, 3rd in GT. Race number: 21. Dumay.
	Coupés de Paris, Montlhéry, DNF. Dumay.
	Tourist Trophy, Goodwood, DNF, crash. Dumay/"Loustel."
	6/9: Coppa Inter Europa Monza, 4th OA. Race number: 59. P. Noblet.
	Sold to Paris.
1966:	Sold to Loye, Paris. Bad crash.
1967:	Damage repaired.
1968:	Sold to M. Mendel.
	Sold to the USA.
1971:	Sold to Delameter.
1972:	Sold to Glen Williams.
1973:	Sold to R. Bodin.
1979:	Sold to Karl Dedolph. Totally restored.
1984:	Entered in Monterey Historic Races.
1990:	Colorado Grand. K. Dedolph.
1994:	For sale.
1995:	Sold to Symbolic Motors.

1523 GT

The last long wheelbase Berlinetta. Delivered to Olivier Gendebien of Belgium in September 1959 in time for the Tour de France. Metal Grey. *Tipo* 508D/ 128 DF-0476DF outside plug engine. Very light conrods. 9 x 33 differential ratio. 40DCL Weber Carburettors. 274bhp at 7400rpm. Originally fitted with disc brakes which were replaced by drum brakes for the 1959 Tour de France as the discs were not homologated. Discs refitted in 1960, together with Koni telescopic dampers. Testa Rossa clutch. Registration number: MO 51647.

1959:	Tour de France, 1st OA. Race number: 163. Gendebien/Bianchi.
	Sold to L. Bianchi/Van der Velde.
	Note: 1033 GT, Gendebien's 1958 Tour de France winner had, at this time, been bought by Lucien Bianchi, crashed at Monza, sent back to the factory for repair and an upright headlight nose. The car was re-numbered as 1523 GT. Was 1523 GT renumbered 1033 GT?
1960:	Coupés USA, Montlhéry. Race number: 18. Bianchi/Van der Velde.
	Auvergne 6 hours, 3rd OA, 2nd in GT. Race number: 78. Bianchi/Van der Velde.
	Laroche Hillclimb, 1st. Van der Velde.
	Anvers 1000km, 1st OA. Race number: 92. Bianchi.
	Paris GP, Montlhéry, 4th OA, 4th in GT. Bianchi/Van der Velde.
	Born Hillclimb, 1st in GT. Van der Velde.
	Andenne Hillclimb, Van der Velde.
	Rouen GP, 5th OA. Bianchi.
	Sold to S. M. Orban.
1961:	Laroche Hillclimb, 1st OA. S. M. Orban.
	Spa 500km, DNF, crash. Race number: 14. Orban. Car destroyed. Engine retrieved and sold in the USA.
1985:	Engine in Texas, USA.

Gear ratios & performance figures

Ratio	1st gear Kph	1st gear Mph	2nd gear Kph	2nd gear Mph	3rd gear Kph	3rd gear Mph	4th gear Kph	4th gear Mph
7 X 32.	86	53.4	122	75.8	164	101.9	202	125.5
8 X 34	93	57.7	131	81.4	176	109.3	217	134.8
8 X 32	99	61.5	136	84.5	184	114.3	231	143.5
9 X 34	104	64.6	148	91.9	199	123.6	245	152.2
9 X 33	107	66.5	152	94.4	205	127.4	252	156.6

With the 7 x 32 back axle ratio installed, factory figures were 0 to 125.5mph in 26 seconds.

Camshaft timing of various 250GT Berlinettas

	Inlet opening advance	Inlet closing retard	Exhaust opening advance	Exhaust closing retard
Early 250GT Competizione	27 degrees	74 degrees	67 degrees	18 degrees
Early 250GT Coupé	22 degrees	66 degrees	67 degrees	17 degrees
1958 250GT Competizione	49 degrees	75 degrees	75 degrees	44 degrees
1958 250GT Competizione	45 degrees	75 degrees	73 degrees	42 degrees
1958 250GT Coupé	28 degrees	73 degrees	68 degrees	17 degrees

Vital statistics

Length:	173 inches
Width:	65 inches
Height:	55 inches
Weight:	2520lbs
Weight, with driver and equipment:	2840lbs
Weight distribution:	49.3/50.7
Power to weight ratio:	10.9lbs per horsepower
Engine displacement:	2953.211cc
Bore and stroke:	73 x 58.8mm
Stroke/bore ratio:	805:1
Typical compression ratio:	9.57:1
Power (typical):	260bhp
Torque:	195lbs ft. At 5,000rpm
Brake area:	136 square inches per ton
Brake lining area:	240 square inches
Dampers:	Houdaille
Steering type:	Worm and sector
Steering wheel turns lock to lock:	3

Acceleration (with 8 X 32 rear axle ratio)

From zero to	Seconds elapsed
30mph	4.0 seconds
40mph	4.8 seconds
50mph	5.8 seconds
60mph	7.8 seconds
70mph	9.3 seconds
80mph	11.1 seconds
90mph	13.7 seconds
100mph	16.1 seconds

Numbers built & chassis numbers

1955-56. Wrap-around rear window, no louvres - 17 cars built.

0369 GT, 0383 GT, 0385 GT, 0393 GT, 0403 GT, 0415 GT, 0425 GT, 0443 GT, 0503 GT, 0507 GT, 0509 GT, 0513 GT, 0539 GT, 0555 GT, 0557 GT, 0563 GT, 0619 GT

1956-57. Fourteen louvres, recessed front headlights - 9 cars built.

0585 GT, 0597 GT, 0607 GT, 0629 GT, 0647 GT, 0677 GT, 0683 GT, 0703 GT, 0707 GT

1957-58. Three slots, covered headlights - 17 cars built.

0723 GT, 0731 GT, 0733 GT, 0747 GT, 0749 GT, 0753 GT, 0763 GT, 0767 GT, 0771 GT, 0773 GT, 0781 GT, 0787 GT, 0793 GT, 0879 GT, 0881 GT, 0893 GT, O895 GT

1958-59 One slot, covered headlights - 39 cars built.

0805 GT, 0897 GT, 0899 GT, 0901 GT, 0903 GT, 0905 GT, 0907 GT, 0909 GT, 0911 GT, 0925 GT, 0931 GT, 0933 GT, 0967 GT, 0969 GT, 0971 GT, 0973 GT, 1031 GT, 0133 GT/1523 GT, 1035 GT, 1037 GT, 0139 GT, 1113 GT, 1127 GT, 1139 GT, 1141 GT, 1143 GT, 1161 GT, 1309 GT, 1321 GT, 1333 GT, 1335 GT, 1353 GT, 1357 GT, 1385 GT, 1389 GT, 1399 GT, 1401 GT

Zagato bodywork - 5 cars built.

0515 GT , 0537 GT, 0665 GT, 0689 GT, 1367 GT

"Interim" model - 7 cars built.

1377 GT, 1461 GT, 1465 GT, 1509 GT, 1521 GT, 1523 GT

Total number built: 94

INDEX

DEAR READER,
WE HOPE YOU HAVE ENJOYED THIS VELOCE PUBLISHING PRODUCTION. IF YOU HAVE IDEAS FOR BOOKS ON FERRARI OR OTHER MARQUES, PLEASE WRITE AND TELL US. MEANTIME, HAPPY MOTORING!